Simulating Fourier Optics Using *Mathematica*®

Simulating Fourier Optics Using *Mathematica*®

Joseph W. Goodman

SPIE PRESS
Bellingham, Washington USA

Library of Congress Control Number: 2024944928

Published by
SPIE
P.O. Box 10
Bellingham, Washington 98227-0010 USA
Phone: +1 360.676.3290
Fax: +1 360.647.1445
Email: books@spie.org
Web: www.spie.org

Printed in the United States of America.
First printing 2024.
For updates to this book, visit http://spie.org and type "PM384" in the search field.

SPIE.

Table of Contents

1. Introduction 1
- 1.1 Some Properties of *Mathematica®* 1
- 1.2 Some Useful Functions and Some Hidden Printing Commands 2
- 1.3 Some Useful Fourier Operators 3

2. From the Wave Equation to the Angular Spectrum 7
- 2.1 The Vector Wave Equation 7
- 2.2 The Scalar Wave Equation 7
- 2.3 The Helmholtz Equation 8
- 2.4 Spatial Fourier Components of a Wave Disturbance as Plane Waves 8
- 2.5 The Angular Spectrum 9
- 2.6 Propagation of the Angular Spectrum 9
- 2.7 The Transfer Function of Free-Space Propagation 11
- 2.8 Diffraction and the Angular Spectrum 12
- 2.9 The Point-Spread Function of Free-Space Propagation 12

3. Fresnel Diffraction 15
- 3.1 Fresnel Approximation to the Transfer Function of Free-Space Propagation 15
- 3.2 Fresnel Approximation to the Point-Spread Function of Free-Space Propagation 15
- 3.3 Equivalence of the Two Fresnel Approximations 16
- 3.4 A Square Aperture and the Convolution Approach 17
- 3.5 A Truncated Gaussian Aperture and the Fresnel Transform Approach 19
- 3.6 A sgn[x] × sgn[y] Function and the Fresnel Transfer Function Approach 21
- 3.7 The Zero-Order Hankel Transform 23
- 3.8 The Projection-Transform Method 24

4. Fraunhofer Diffraction Patterns 29
- 4.1 Square Aperture 29
- 4.2 Circular Aperture 30
- 4.3 When is the Fraunhofer Approximation Accurate? 31

5. Diffraction by an Arbitrary Linear Optical System 33
- 5.1 Ray-Transfer (or ABCD) Matrices 33
- 5.2 Wave Analysis of Complex Optical Systems Using Ray-Transfer Matrices 35

6. Discrete Diffraction Calculations 41
- 6.1 The Discrete Fourier Transform (DFT) 41
- 6.2 Discrete Diffraction Equations 42
- 6.3 Sampling Requirements 45
- 6.4 Some Discrete Diffraction Examples With *Mathematica* 47
- 6.5 A Diffraction Pattern That Does Not Change Shape as the Fresnel Number Changes 52

7. Imaging With Coherent Light 55
- 7.1 Coherent vs. Incoherent Imaging 55

7.2 The Amplitude Impulse Response and Transfer Function 55
7.3 Coherent Imaging System Model 56
7.4 Symbolic Simulation Example 57
7.5 Discrete Imaging Examples 61
7.6 Coherent Image Speckle 66
8. Imaging With Incoherent Light 69
8.1 The Intensity Impulse Response and the Optical Transfer Function (OTF) 69
8.2 The Object 69
8.3 The Focal-Plane Aperture 70
8.4 The Optical Transfer Function 71
8.5 Finding the Incoherent Image 71
9. Imaging With Holography 73
9.1 Gabor Holography 73
9.2 Leith–Upatnieks Holography – The Fourier Hologram 78
9.3 Leith–Upatnieks Holography – The Fresnel Hologram 81
9.4 Phase-Stepping Holography 87
10. Spatial Filtering With Coherent Optics 91
10.1 An Object to be Filtered 91
10.2 Filtering to Emphasize Horizontal or Vertical Bars 92
10.3 Edge Enhancement 94
10.4 Restoring Resolution in an Incoherent Blurred Image 95
10.5 An Inverse Filter 96
10.6 The Wiener Filter 98

Acknowledgment 103
References 103

1. Introduction

Fourier optics and *Mathematica*® individually are subjects about which textbooks already exist. However, a textbook that combines both topics with an emphasis on simulation is new. In this book, we use *Mathematica* to illustrate many of the concepts of Fourier optics, hopefully providing the reader insights into both subjects.

We assume at the start that the reader has a good understanding of Fourier transforms, such as can be provided by the ebook *Fourier Transforms Using* Mathematica® published by SPIE Press in 2020 [1], or by several older and more detailed texts [2], [3]. Knowledge of *Mathematica* is helpful but not required, since the reader can learn this language, at least at the elementary level, by following the commands introduced in this text. A far more complete treatment of Fourier optics can be found in Ref. [4].

1.1 Some Properties of *Mathematica*®

One of the beauties of *Mathematica* is that some figures can be produced in an interactive form, for which the reader is able change the plotting parameters and see the changes of the calculated results. Such capability cannot be duplicated in a printed version of this book. However, the reader can access the *Mathematica* files at the link https://spie.org/Samples/Pressbook_Supplemental/PM384_sup.zip. In some cases, manipulable figures are included in hidden cells that the user of the code can un-hide. To have this capability the book must be read with either the full *Mathematica* program or the free software package Wolfram Player, available from the Wolfram website. In principle, this book can also be read with a web browser if the book is stored in the Wolfram Cloud, but this option does not work well for large file sizes.

An attempt has been made to provide enough introductory *Mathematica* code to allow a student to learn the relevant commands on the fly. In what follows we present some salient features of *Mathematica* (repeated from *FourierTransforms Using Mathematica* [1]) that will help the novice get started.

1. Built-in functions in *Mathematica* always begin with upper-case letters. If the command consists of a concatenation of two separate words, both words must begin with upper-case letters.

2. In order to avoid confusion with built-in commands, user-defined commands should usually begin with a lower-case letter. However, it is permissible to begin commands with an upper-case letter in a font that is different than Source Code Pro, the usual font for *Mathematica* input. As an example, $\mathcal{N}$ is not the same as `N`.

3. *Mathematica* reduces an input that is a rational number to the equivalent simplest rational number possible. The ratio `32/6` **yields** `16/3` but does not produce an approximate decimal result. To obtain a decimal result, place a decimal point at the end of either the numerator or the denominator. Thus, `32./6` yields the result `5.33333`.

4. The arguments of functions must be enclosed in square brackets `[]`. Curly brackets `{}` are reserved for containing lists, which can represent vectors and matrices, and are also used for variable ranges in, for example, plot commands. Parentheses are reserved for grouping mathematics.

5. When a function is first defined, each independent variable on the left of the definition must be followed by an underbar `_`. Underbars are not used on the right of the definition or when the function is called in later code. The underbar indicates that the argument is a pattern-matching construct, and that the argument should not be interpreted as a specific variable. Any independent variable can be used in future calls of the function.

6. The symbol `:=` means delayed inequality, for which the operation of equality is made only when the symbol on the left is called in later programming.

7. The symbol `%`, when used as the argument of a function, represents the last previous output, which is now used as the input to this new function. `%%` represents the second-to-last output, etc.

8. Often certain symbols may be used more than once in a program, with their meaning having changed between uses. If `h` is the symbol, then the command `Clear[h]` erases the definition of the symbol `h` and allows a new definition to be made. To clear more than one symbol, for example, `h` and `g`, use `Clear[h,g]`. To clear all symbols and functions that have been previously defined, use the command `Clear["Global`*"]`.

9. The symbol `;` at the end of an expression to be evaluated indicates that the evaluation should take place, but the result of the evaluation should be suppressed until it is needed later in the program.

10. The symbol `I` (or `ⅈ`) is used in *Mathematica* for the imaginary constant $\sqrt{-1}$. The symbol `D[]` indicates a derivative operation.

11. To execute a command in *Mathematica*, place the cursor in the same cell (cells are indicated by vertical lines on the right of the notebook) as the command, and press the shift key and the return key simultaneously.

12. The symbols `&&` mean a logical *and*; the symbols `||` mean a logical *or*.

13. Multiplication can be indicated by either `*` or by a space between the two quantities to be multiplied.

14. When dealing with a complex number `c`, the command `Abs[c]` yields the magnitude of `c`, and `Arg[c]` gives the phase angle of `c` in radians.

15. *Mathematica* input commands are always written in Source Code Pro semi-bold font. In the event that a *Mathematica* command extends beyond the end of a line, you may find that a hyphen has been inserted in the command. This hyphen is not part of the command. When viewing this material with either the *Mathematica* or the Wolfram Player programs, if you change the size of the window, the line breaks will change, with the result that these hyphens should disappear.

The author wishes to emphasize that he is not an "expert" *Mathematica* user, but has mastered the program well enough to perform almost any calculation he wishes. Hopefully, the reader will arrive at a similar level after reading this book. But with *Mathematica*, one can always dive deeper and learn new tricks. That's part of the fun of using the program.

1.2 Some Useful Functions and Some Hidden Printing Commands

There are certain functions that occur frequently enough in Fourier optics that defining special notations for them saves time. These user-defined functions can be expressed in *Mathematica* equivalents, as the list below shows. User-defined functions are on the left, and equivalent *Mathematica* expressions are on the right.

In[ ]:=

```
rect[x_] := UnitBox[x];
sinc[x_] = Sinc[Pi * x];
step[x_] := UnitStep[x];
sgn[x_] := Sign[x];
gaus[x_] := Exp[-Pi * x^2];
jinc[x_] := 2 * BesselJ[1, Pi * x] / (Pi * x);
tri[x_] := UnitTriangle[x];
rtri[x_] := UnitBox[x - 1 / 2] * UnitTriangle[x]
δ[x_] := DiracDelta[x]
comb[x_] := DiracComb[x]
circ[r_] := UnitBox[r - 1 / 2]
jinc[r_] := 2 * BesselJ[1, Pi * r] / (Pi * r)
chat[r_] := (2 / Pi) * (ArcCos[r] - r * Sqrt[1 - r^2]) * UnitBox[r - 1 / 2]
gaus[r_] := Exp[-Pi * r^2]
```

Note that the last four functions listed are functions of radius r and are therefore usually encountered in describing 2D functions. Additional 2D functions can be composed as products of two 1D functions, one depending on x and the other on y. The fact that the cell containing these definitions has a gray background is *Mathematica*'s indication that the cell has been defined as an initialization cell, which will be evaluated at the beginning of any evaluation of this chapter. In later chapters we insert a hidden initialization cell that provides definitions of all user-defined functions needed in each chapter. In this way, the reader with access to the software can evaluate chapters individually.

An additional initialization cell that has been hidden from view has commands defining parameters of plots that will hold in each chapter. If you wish to see these commands in this chapter, click on the small cell sign at the right end of the gray line below the following sentence, then go to the *Mathematica* menu at the top of your screen, click on "Cell," then on "Cell Properties," then on "Open." The hidden cell will then open up for you to view.

1.3 Some Useful Fourier Operators

We will have many occasions to perform Fourier transforms and inverse transforms. These occasions sometimes involve symbolic functions, and sometimes discrete data. We therefore define both 1D and 2D symbolic Fourier transforms and their inverses, as well the discrete Fourier transform and its inverse. In the case of the discrete Fourier transform (DFT), the same *Mathematica* command will work for both 1D and 2D data (or discrete data of any dimension). In defining these commands, we have chosen Fourier parameters that yield the form of the Fourier transform that we will use in the sections and chapters that follow. Note that we have chosen to have a negative exponential kernel for the forward Fourier transform and a positive exponential kernel for the inverse Fourier transform, as was done in Refs. [2] and [4]. *Mathematica* defines the symbolic Fourier transform with two parameters, **a** and **b**. In one dimension, the Fourier transform $\mathcal{G}(\nu)$ of a function $\mathbf{g}(x)$ is defined as

$$\mathcal{G}(\nu) = \sqrt{\frac{|b|}{(2\pi)^{1-a}}} \int_{-\infty}^{\infty} \mathbf{g}(x) \exp(i\, b\, \nu\, x)\, dx,$$

and the inverse Fourier transform as

$$\mathbf{g}(x) = \sqrt{\frac{|b|}{(2\pi)^{1+a}}} \int_{-\infty}^{\infty} \mathcal{G}(\nu) \exp(-i\, b\, \nu\, x)\, d\nu\,.$$

For our definition of the Fourier transform, **a** = 0 and **b** = -2π in both the forward and inverse transforms. Thus, the operator forms of the 1D and 2D symbolic forward Fourier transforms are as follows (presented as initialization cells):

```
In[ ]:= ft1[arg_] := FullSimplify[
          FourierTransform[arg, x, ν, FourierParameters → {0, -2 π}]]
```

We have included the **FullSimplify[]** command to get the simplest form of the result that *Mathematica* can produce. Here *arg* represents the function of *x* to be transformed.

The 1D inverse symbolic Fourier transform operator is

```
In[ ]:= ift1[arg_] := FullSimplify[InverseFourierTransform[
          arg, ν, x, FourierParameters → {0, -2 π}]]
```

In a similar fashion, the 2D forward Fourier transform can be represented by an operator:

```
In[ ]:= ft2[arg_] := FullSimplify[FourierTransform[
          arg, {x, y}, {νx, νy}, FourierParameters → {0, -2 π}]]
```

The 2D inverse symbolic Fourier transform is

```
In[ ]:= ift2[arg_] := FullSimplify[InverseFourierTransform[
          arg, {νx, νy}, {x, y}, FourierParameters → {0, -2 π}]]
```

The forward *discrete* Fourier transform for both 1D and 2D, where $\mathbf{g}_n$ is a discrete array, is defined by *Mathematica* to be

$$\mathcal{G}_m = \frac{1}{N^{(1-a)/2}} \sum_{n=1}^{N} \mathbf{g}_n\, e^{i \frac{2\pi}{N} b (n-1)(m-1)} \qquad m = 1, \cdots, N,$$

where **a** and **b** are parameters and *Mathematica* numbers all sequences starting with index 1. $\mathbf{g}_n$ and $\mathcal{G}_m$ are both sequences (called "lists" in *Mathematica*) of length *N*. The parameters **a** and **b** are to be chosen. For a conventional DFT as used in signal processing, we would usually choose (**a**, **b**) = (1, –1) (again we use a negative exponent to represent a forward transform). Thus, an operator for the discrete Fourier transform of numerical data can be written as

```
In[∘ ]:= dft[arg_] :=
  FullSimplify[Fourier[arg, FourierParameters → {1,-1}]]
```

where *arg* is the sequence to be transformed.

The inverse discrete Fourier transform for both 1D and 2D is defined in *Mathematica* as

$$\mathbf{g}_n = \frac{1}{N^{(1+a)/2}} \sum_{m=1}^{N} \sum_{n=1_m}^{N} \mathcal{G}_m \; e^{-i \frac{2\pi}{N} b(m-1)(n-1)} \qquad n = 1, \cdots, N.$$

The parameters for the inverse DFT are again (**a**, **b**) = (1, –1). Thus, the operator representation for the inverse DFT is

```
In[∘ ]:= idft[arg_] :=
  FullSimplify [InverseFourier [arg, FourierParameters → {1,-1}]]
```

When any of these user-defined operators are used in future chapters, they will be defined again at the beginning of the chapter in a hidden cell so that the chapters can be evaluated independently.

2. From the Wave Equation to the Angular Spectrum

Virtually all properties of optical waves in linear media can be derived from the vector wave equation. In this chapter, we start with that equation and gradually simplify it under various assumptions or restrictions. In addition, we introduce the angular spectrum and its properties. An alternative discussion of this material is found in Ref. [4].

2.1 The Vector Wave Equation

Maxwell's equations are the most fundamental starting point for deriving properties of optical waves. Using these equations, we can derive the vector wave equation that governs propagation of electromagnetic disturbances in all linear, isotropic, homogeneous, non-dispersive, non-magnetic, and source-free media:

$$\nabla^2 \vec{\mathcal{E}} - \frac{n^2}{c^2}\frac{\partial^2}{\partial t^2}\vec{\mathcal{E}} = 0,$$

where $\vec{\mathcal{E}}$ represents the vector electric field, n is the refractive index, c is the vacuum speed of light, t represents time, and ∇^2 is the Laplacian operator, $\nabla^2 = \frac{\partial^2}{\partial x^2} + \frac{\partial^2}{\partial y^2} + \frac{\partial^2}{\partial z^2}$. The same partial differential equation holds for the vector magnetic field $\vec{\mathcal{H}}$.

2.2 The Scalar Wave Equation

Fortunately, the wave equation can be approximated in a number of ways that lead us to simpler mathematical forms to deal with. Foremost among those approximations is a transition from a vector equation to a scalar equation. In free space, the x, y, and z components of $\vec{\mathcal{E}}$ and $\vec{\mathcal{H}}$ each obey a scalar version of the vector wave equation; for example,

$$\nabla^2 \mathcal{E}_X - \frac{n^2}{c^2}\frac{\partial^2}{\partial t^2}\mathcal{E}_X = 0,\ \ \nabla^2 \mathcal{E}_Y - \frac{n^2}{c^2}\frac{\partial^2}{\partial t^2}\mathcal{E}_Y = 0,\ \ \nabla^2 \mathcal{H}_X - \frac{n^2}{c^2}\frac{\partial^2}{\partial t^2}\mathcal{H}_X = 0,\quad \text{etc.}$$

These vector components remain uncoupled in free space, and therefore their behavior can be solved for independently.

However, there are conditions under which coupling between the components of $\vec{\mathcal{E}}$ and the components of $\vec{\mathcal{H}}$ can exist. This happens at boundaries where material objects interact with the propagating waves. The boundary conditions on such objects can impose relations between the components of $\vec{\mathcal{E}}$ and $\vec{\mathcal{H}}$.

Fortunately, when the optical wave is propagating in free space, there are no couplings between components, and the scalar wave equation can be used to derive accurate results. In addition, there are approximations to the boundary conditions that ignore coupling and still yield reasonably accurate results. Because all field components obey the same scalar wave equation, we can represent any one of them by a scalar function $u(P, t)$, where P represents a position in space with coordinates (x, y, z) and t represents time. Thus, the scalar wave equation we will work with in the future is

$$\nabla^2 u(P,t) - \frac{n^2}{c^2}\frac{\partial^2}{\partial t^2}u(P,t) = 0.$$

2.3 The Helmholtz Equation

When the optical disturbance is monochromatic with frequency ν, a simplification of the scalar wave equation occurs. In such a case, the scalar wave u can be assumed to be of the form

$$u(P, t) = \mathcal{A}(P)\cos[2\pi\nu t - \phi(P)],$$

where ν is the frequency of the wave, $\mathcal{A}$ is its magnitude, and ϕ is its phase. It is convenient to represent the real-valued co-sinusoidal waveform $u(P, t)$ by a complex-valued analytic signal. We represent all complex quantities by bold-faced, non-italic characters; thus,

$$\mathbf{u}(P, t) = \mathbf{U}(P)\exp(-i2\pi\nu t),$$

where $\mathbf{U}(P) = \mathcal{A}(P)\exp[i\phi(P)]$ is called the complex amplitude of the monochromatic wave. Note that we can always recover the real-valued signal by taking the real part of the analytic signal $\mathbf{u}(P, t)$.

We can use *Mathematica*® to simplify the second term of the wave equation. Referring to the scalar wave equation above, we calculate $-n^2/c^2$ times the second partial derivative of $\mathbf{u}$ with respect to t using *Mathematica*:

```
In[]:= -(n^2 / c^2)*D[U[P]*Exp[-I*2*Pi*ν*t],{t, 2}]
```

$$\frac{4\, e^{-2 i \pi t \nu}\, n^2 \pi^2 \nu^2\, U[P]}{c^2}$$

Note that the command `D[function,{t,n}]` performs an nth-order partial derivative of the function with respect to the variable t. Now we define the scalar wavenumber $k = 2\pi/\lambda$, where λ is the wavelength and $\nu\,\lambda = c/n$, with c being the vacuum speed of light and n the refractive index of the medium within which the light is propagating. It follows that $4\pi^2 n^2\nu^2/c^2 = 4\pi^2/\lambda^2 = k^2$, and we conclude that for monochromatic light, the scalar wave equation can be reduced to

$$\left(\nabla^2 + k^2\right)\mathbf{U}(P) = 0.$$

This result is known as the Helmholtz equation and serves as the starting point for simplifying diffraction problems.

Our goal now is to progress further towards establishing a framework within which it will be possible to calculate the behavior of light as it propagates through free space and as it encounters apertures through which it must pass. Our approach will be to decompose the complex field amplitude $\mathbf{U}(x, y, z)$ into Fourier components, and to identify those components as representations of plane waves propagating with various angles with respect to the optical axis.

2.4 Spatial Fourier Components of a Wave Disturbance as Plane Waves

Consider the distribution of complex amplitude $\mathbf{U}(x, y; 0)$ in a plane located at axial coordinate $z = 0$. That distribution can be represented by an inverse Fourier transform of its Fourier spectrum $\mathcal{U}(\nu_x, \nu_y; 0)$:

$$\mathbf{U}(x, y; 0) = \iint_{-\infty}^{\infty} \mathcal{U}(\nu_x, \nu_y; 0)\, e^{i 2\pi\left(\nu_x x + \nu_y y\right)}\, d\nu_x\, d\nu_y.$$

This representation can be regarded as an infinite sum of complex exponential elementary signals, each with a different spatial frequency pair (ν_x, ν_y) and weighted by a complex amplitude $\mathcal{U}(\nu_x, \nu_y; 0)$. The complex exponentials have a physical interpretation as plane waves traveling in different directions, as we will now see.

Consider a unit-amplitude plane wave traveling with a wave vector $\vec{k} = k_x\hat{x} + k_y\hat{y} + k_z\hat{z}$ pointing in the direction of propagation, where $(\hat{x}, \hat{y}, \hat{z})$ are unit vectors pointing in the x, y, and z directions. The mathematical representation of such a plane wave would be

$$\mathbf{p}(x, y, z, t) = \exp\left[i\left(\vec{k} \cdot \vec{r} - 2\pi\nu t\right)\right] = \exp[i(k_x x + k_y y + k_z z)]\exp(-i2\pi\nu t),$$

where $\vec{r}$ is a vector pointing from the origin to the point (x, y, z), $|\vec{k}| = 2\pi/\lambda$, and the $\cdot$ represents a vector dot product. Suppressing the time dependence and focusing on the plane $z = 0$, the complex amplitude of the plane wave takes the form

$$\mathbf{P}(x, y; 0) = \exp[i(k_x x + k_y y)].$$

The similarity between the complex exponential elementary functions in the Fourier decomposition and the plane wave complex amplitude is clear.

To explore the relationship further, let the direction of the k-vector be represented by its direction cosines, (α, β, γ), where α is the cosine of the angle between the k-vector and the x axis, β is the cosine of the angle between the k-vector and the y axis, and γ is the cosine of the angle between the k-vector and the z axis. Note for future use that $\gamma = \sqrt{\alpha^2 + \beta^2}$. The complex amplitude of the plane wave in the $z = 0$ plane can be written as

$$\mathbf{P}(x, y; 0) = \exp\left[i2\pi\left(\frac{\alpha}{\lambda}x + \frac{\beta}{\lambda}y\right)\right],$$

and we see that the spatial frequencies in the Fourier decomposition correspond to direction cosines of the angles of propagation, normalized by the wavelength λ. That is, $(\nu_x, \nu_y) = \left(\frac{\alpha}{\lambda}, \frac{\beta}{\lambda}\right)$.

2.5 The Angular Spectrum

Drawing on the results of the previous section, we define the angular spectrum of plane waves $\mathbf{A}\left(\frac{\alpha}{\lambda}, \frac{\beta}{\lambda}; 0\right)$ in the plane $z = 0$ by the Fourier transform relationship

$$\mathbf{A}\left(\frac{\alpha}{\lambda}, \frac{\beta}{\lambda}; 0\right) = \int\int_{-\infty}^{\infty} \mathbf{U}(x, y; 0)\, e^{-i2\pi\left(\frac{\alpha}{\lambda}x + \frac{\beta}{\lambda}y\right)}\, dx\, dy.$$

With this definition comes the inverse Fourier transform relationship:

$$\mathbf{U}(x, y; 0) = \int\int_{-\infty}^{\infty} \mathbf{A}\left(\frac{\alpha}{\lambda}, \frac{\beta}{\lambda}; 0\right) e^{i2\pi\left(\frac{\alpha}{\lambda}x + \frac{\beta}{\lambda}y\right)}\, d\frac{\alpha}{\lambda}\, d\frac{\beta}{\lambda}.$$

The latter equation is a representation of $\mathbf{U}(x, y; 0)$ as a superposition of plane waves, each plane wave weighted by a complex-valued weighting factor $\mathbf{A}\left(\frac{\alpha}{\lambda}, \frac{\beta}{\lambda}; 0\right)$. When added together, the plane waves yield the complex field $\mathbf{U}(x, y; 0)$ in the plane $z = 0$. There is a subtlety to be addressed shortly. An inverse Fourier transform can represent any reasonably well-behaved function with a doubly infinite combination of complex exponentials, with spatial frequencies extending to $\pm\infty$ if necessary. However, α and β, being direction cosines of plane waves, can never exceed unity. Therefore, we need to address the nature of the Fourier components when the limits of integration extend beyond $1/\lambda$ or $-1/\lambda$. We shall do so in the next section.

2.6 Propagation of the Angular Spectrum

A plane wave propagating in free space with k-vector $\vec{k}$ remains a plane wave with the same k-vector throughout free space. Furthermore, the complex amplitude of that plane wave does not change with propagation through free space. Only by interacting with material structures does the plane-wave complex amplitude change.

However, plane waves propagating in different directions travel different distances to reach the origin of a perpendicular plane at axial coordinate z. The result is that plane waves with different directions of propagation suffer relative phase shifts, and it is because of these phase shifts that the structure of the complex field $\mathbf{U}(x, y, z)$ changes with propagation distance z. We will formalized these changes with the mathematics that follow.

Recall that the complex amplitude $\mathbf{U}$ of a monochromatic wave disturbance must satisfy the Helmholtz equation,

$$\left(\nabla^2 + k^2\right)\mathbf{U}(P) = 0.$$

In addition, at axial distance z, the complex field is given by

$$\mathbf{U}(x, y; z) = \int\int_{-\infty}^{\infty} \mathbf{A}\left(\frac{\alpha}{\lambda}, \frac{\beta}{\lambda}; z\right) e^{i2\pi\left(\frac{\alpha}{\lambda}x + \frac{\beta}{\lambda}y\right)} d\frac{\alpha}{\lambda}\, d\frac{\beta}{\lambda}.$$

Applying the Helmholtz equation directly to this representation of $\mathbf{U}$, we find that in any plane at distance z, the angular spectrum $\mathbf{A}(\alpha/\lambda, \beta/\lambda; z)$ must satisfy the differential equation

$$\frac{d^2}{dz^2}\mathbf{A}(\alpha/\lambda, \beta/\lambda; z) + \left(\frac{2\pi}{\lambda}\right)^2\left[1 - \alpha^2 - \beta^2\right]\mathbf{A}(\alpha/\lambda, \beta/\lambda; z) = 0.$$

This differential equation is a relatively easy one to solve, but since this book is introducing the reader to *Mathematica* as well as Fourier optics, we choose to use the *Mathematica* command **`DSolve[]`** to obtain a solution. The format of the **`DSolve[]`** command can be explained if we write it as **`DSolve[1,2,3]`**, where 1, 2, and 3 are different parts of the argument separated by commas. Part 1 consists of the differential equation itself, using the command **`D[]`** to represent derivatives. Part 2 is the quantity to be solved for, in this case, $\mathbf{A}(\alpha/\lambda, \beta/\lambda; z)$. Part 3 is the independent variable of interest, in this case, z. Thus, to find the possible solutions to the above equation, we write and evaluate the following code:

```
In[ ]:= DSolve[D[A[α / λ, β / λ, z], {z, 2}] +
          (2 π / λ) ^2 (1 - α^2 - β^2) * A[α / λ, β / λ, z] == 0, A[α / λ, β / λ, z], z]
```

Out[ ]=

$$\left\{\left\{A\left[\frac{\alpha}{\lambda}, \frac{\beta}{\lambda}, z\right] \to e^{\frac{2\pi z\sqrt{-1+\alpha^2+\beta^2}}{\lambda}} c_1 + e^{-\frac{2\pi z\sqrt{-1+\alpha^2+\beta^2}}{\lambda}} c_2\right\}\right\}$$

We see that *Mathematica* produces two possible solutions, each with an unknown multiplier to be determined by boundary conditions. To explore these solutions further, factor $i = \sqrt{-1}$ out of each of the square roots in the exponentials, yielding

$$\mathbf{A}(\alpha/\lambda, \beta/\lambda; z) = \mathbf{c}_1\, e^{i\frac{2\pi z}{\lambda}\sqrt{1-\alpha^2-\beta^2}} + \mathbf{c}_2\, e^{-i\frac{2\pi z}{\lambda}\sqrt{1-\alpha^2-\beta^2}}.$$

Consider what happens when $\alpha^2 + \beta^2 > 1$. We can then factor another $\sqrt{-1}$ out of the square root, yielding

$$\mathbf{A}(\alpha/\lambda, \beta/\lambda; z) = \mathbf{c}_1\, e^{-\frac{2\pi z}{\lambda}\sqrt{\alpha^2+\beta^2-1}} + \mathbf{c}_2\, e^{\frac{2\pi z}{\lambda}\sqrt{\alpha^2+\beta^2-1}}.$$

One of these two solutions can be ruled out on physical grounds. The first solution contains an exponential that attenuates as z increases in the positive direction, while the second solution contains an exponential that grows without bound as z increases in the positive direction. Physically, only the first term is allowable for positive z. The second term can be thought of as representing a backward propagating wave that attenuates with increasingly negative z. Given that the first term is the desired solution, what is the constant $\mathbf{c}_1$? This constant is determined by the initial condition,

$$\lim_{z\to 0} \mathbf{A}(\alpha/\lambda,\ \beta/\lambda;\ z) = \mathbf{A}(\alpha/\lambda,\ \beta/\lambda;\ 0).$$

We conclude that propagation of the angular spectrum in the positive z direction in free space is governed entirely by the equation

$$\mathbf{A}(\alpha/\lambda,\ \beta/\lambda;\ z) = \mathbf{A}(\alpha/\lambda,\ \beta/\lambda;\ 0)\ e^{i\frac{2\pi z}{\lambda}\sqrt{1-\alpha^2-\beta^2}}.$$

Thus, the initial angular spectrum $\mathbf{A}(\alpha/\lambda, \beta/\lambda; 0)$, determined by the boundary condition in the $z = 0$ plane, is modified by propagation by a pure phase dispersion when $\alpha^2 + \beta^2 \leq 1$. On the other hand, when $\alpha^2 + \beta^2 > 1$, the components of the angular spectrum satisfying this inequality are exponentially attenuated. We return to this latter case in a moment.

2.7 The Transfer Function of Free-Space Propagation

Wave propagation in free space is a linear phenomenon, as evidenced by the linearity of the wave equation. We see that it is also an invariant linear phenomenon, by virtue of the fact that the propagation phenomenon can be characterized by a transfer function. Remembering that $\nu_x = \alpha/\lambda$ and $\nu_y = \beta/\lambda$, we can write that the Fourier spectrum $\mathcal{U}(\nu_x, \nu_y; z)$ of $\mathbf{U}(x, y, z)$ is related to the Fourier spectrum $\mathcal{U}(\nu_x, \nu_y; 0)$ in the plane $z = 0$ by

$$\mathcal{U}(\nu_x, \nu_y; z) = \mathcal{U}(\nu_x, \nu_y; 0)\, e^{i\frac{2\pi z}{\lambda}\sqrt{1-(\lambda\nu_x)^2-(\lambda\nu_y)^2}}.$$

The transfer function of free-space propagation is therefore

$$\mathcal{H}(\nu_x, \nu_y; z) = \frac{\mathcal{U}(\nu_x, \nu_y; z)}{\mathcal{U}(\nu_x, \nu_y; 0)} = e^{i\frac{2\pi z}{\lambda}\sqrt{1-(\lambda\nu_x)^2-(\lambda\nu_y)^2}}.$$

The transfer function is seen to be a pure phase dispersion for $\nu_x^2 + \nu_y^2 \leq 1/\lambda^2$.

Note that this transfer function is circularly symmetric in the (ν_x, ν_y) plane. If we define radius ρ in the frequency plane as $\rho = \sqrt{\nu_x^2 + \nu_y^2}$, the transfer function can also be written as

$$\mathcal{H}(\rho; z) = e^{i\frac{2\pi z}{\lambda}\sqrt{1-(\lambda\rho)^2}}.$$

The behavior of the transfer function when $\nu_x^2 + \nu_y^2 \geq 1/\lambda^2$ requires more discussion. It is quite possible that the boundary condition at $z = 0$ will result in Fourier components that extend beyond the limit of $1/\lambda$ along one or both axes in the frequency plane. For example, a simple circular aperture has frequency components that extend to infinity in all directions. As implied by the transfer function, frequencies higher than $1/\lambda$ will suffer pure attenuation and will die away exponentially with increases in propagation distance z. Such components of the angular spectrum are called *evanescent waves*. They do not carry energy in the z direction and in most cases can be ignored when calculating the field distribution at distances z that are at least several wavelengths.

It is possible to couple to evanescent waves, effectively changing them to propagating waves. This can be accomplished either by placing very high-spatial-frequency structures in very close proximity to the $z = 0$ plane, or by illuminating the object at a very steep angle.

We turn now to considering the effects of limiting apertures or other structures in the $z = 0$ plane on the angular spectrum transmitted beyond that plane.

2.8 Diffraction and the Angular Spectrum

Let the complex field incident on the $z = 0$ plane again be represented by $\mathbf{U}(x, y; 0^-)$ and let the complex field leaving the $z = 0$ plane be represented by $\mathbf{U}(x, y; 0^+)$. Then the complex-valued amplitude transmittance $\mathbf{t}_A(x, y)$ through the $z = 0$ plane can be defined by the ratio of $\mathbf{U}(x, y; 0^+)$ to $\mathbf{U}(x, y; 0^-)$, or equivalently by

$$\mathbf{U}(x, y; 0^+) = \mathbf{t}_A(x, y)\, \mathbf{U}(x, y; 0^-).$$

The amplitude transmittance $\mathbf{t}_A(x, y)$ can represent an attenuation at each (x, y), a phase shift at each (x, y), or a combination of both.

What is the effect of the amplitude transmittance in the $z = 0$ plane on the angular spectrum that is transmitted? The convolution theorem of Fourier transform analysis implies that the angular spectrum transmitted at the $z = 0^+$ plane is the convolution of the angular spectrum incident on that plane with the angular spectrum of the amplitude transmittance. Thus,

$$\mathbf{A}\left(\frac{\alpha}{\lambda}, \frac{\beta}{\lambda}; 0^+\right) = \int\int_{-\infty}^{\infty} \mathbf{A}\left(\frac{\xi}{\lambda}, \frac{\eta}{\lambda}; 0^-\right) \mathbf{T}_A\left(\frac{\alpha-\xi}{\lambda}, \frac{\beta-\eta}{\lambda}\right) d\frac{\xi}{\lambda}\, d\frac{\eta}{\lambda} = \mathbf{A}\left(\frac{\alpha}{\lambda}, \frac{\beta}{\lambda}; 0^-\right) \otimes \mathbf{T}_A\left(\frac{\alpha}{\lambda}, \frac{\beta}{\lambda}\right),$$

where the symbol $\otimes$ represents a convolution. Often the transmitting structure in the $z = 0$ plane is an aperture, and the finite size of the aperture has the effect of broadening the angular spectrum. However, in some cases, such as when a lens is placed in the $z = 0$ plane, the angular spectrum may be compressed rather than broadened.

Thus, the diffraction phenomenon can be envisioned first as a broadening or narrowing of the angular spectrum as the incident wave passes through the $z = 0$ plane, and then as a phase dispersion of the transmitted angular spectrum as it propagates to larger distances. Because the amount of phase dispersion depends on how far the wave propagates, the distribution of field strength changes as propagation takes place.

2.9 The Point-Spread Function of Free-Space Propagation

We have seen that propagation through free space can be characterized by a 2D transfer function. This implies that there exists a 2D complex-valued point-spread function (impulse response) that is the inverse Fourier transform of the transfer function. Unfortunately, the form of the transfer function is sufficiently complex that *Mathematica* is unable to find a symbolic solution for the inverse transform. We therefore rely on work by Sherman [5], who found the required Fourier transform pair by starting with the point-spread function derived by a different method, and showing that its corresponding transfer function is indeed the function we have found here. The form of the point-spread function, which you can also find derived in Ref. [4] from first principles, is

$$\mathbf{h}(x, y; z) = \frac{1}{2\pi}\frac{z}{r}\left(\frac{1}{r} - i\,k\right)\frac{\exp(i\,k\,r)}{r},$$

where $r = \sqrt{x^2 + y^2 + z^2}$ and again $k = 2\pi/\lambda$. Like the transfer function, the point-spread function is circularly symmetric in the (x, y) plane at any distance z. Knowing the form of the point-spread function, we can in principle express the complex field at distance z as a convolution of the field $\mathbf{U}(x, y; 0^+)$ in the plane $z = 0^+$ with $\mathbf{h}(x, y; z)$:

$$\mathbf{U}(x, y; z) = \mathbf{U}(x, y; 0^+) \otimes \mathbf{h}(x, y; z).$$

Symbolic computations using either the point-spread function or the transfer function in the forms expressed above generally fail to produce a symbolic answer (with a few exceptions). Approximations and simplifications are needed, and we turn our attention to them in the following chapter.

3. Fresnel Diffraction

In this chapter, we explore approximations to the transfer function and point-spread function of free-space propagation in order to find analytical forms that are amenable to symbolic results. We explore an approximation to the transfer function first, then an approximation to the point-spread function. We then show that the two approximations are in fact the same. Again, a more detailed discussion of the background can be found in Ref. [4].

3.1 Fresnel Approximation to the Transfer Function of Free-Space Propagation

Consider again the form of the transfer function of free-space propagation:

$$\mathcal{H}(\nu_x, \nu_y; z) = e^{i\frac{2\pi z}{\lambda}\sqrt{1-(\lambda\nu_x)^2-(\lambda\nu_y)^2}}.$$

Suppose that we restrict attention to diffracting objects that generate spatial frequency components ν_x and ν_y that are much less than $1/\lambda$. Equivalently, we consider only components of the angular spectrum for which α and β are much less than 1. Such components correspond to plane waves that are propagating with k-vectors that are close to the optical axis (i.e., the z axis). Under such a circumstance, we can approximate the square root in the exponent as

$$\sqrt{1-(\lambda\nu_x)^2-(\lambda\nu_y)^2} \approx 1 - \frac{(\lambda\nu_x)^2}{2} - \frac{(\lambda\nu_y)^2}{2}.$$

The form of the transfer function then becomes

$$\mathcal{H}(\nu_x, \nu_y; z) \approx e^{i\frac{2\pi z}{\lambda}} e^{-i\pi\lambda z\left(\nu_x^2+\nu_y^2\right)}.$$

The first exponential factor in this expression is a constant phase delay across the (ν_x, ν_y) plane and can be dropped since it does not affect the form of the result. We accordingly define the Fresnel transfer function by

$$\mathcal{H}_F(\nu_x, \nu_y; z) = e^{-i\pi\lambda z\left(\nu_x^2+\nu_y^2\right)}.$$

This is a considerably simpler form of the transfer function than the original, and, as we shall see, allows many symbolic diffraction results to be obtained.

To find the complex field **U**$(x, y; z)$ given the complex field **U**$(x, y; 0^+)$, we simply Fourier transform **U**$(x, y; 0^+)$, multiply that transform by $\mathcal{H}_F(\nu_x, \nu_y; z)$, and inverse Fourier transform the product.

3.2 Fresnel Approximation to the Point-Spread Function of Free-Space Propagation

As a starting point, again consider the form of the point-spread function as stated in the previous chapter:

$$\mathbf{h}(x, y; z) = \frac{1}{2\pi}\frac{z}{r}\left(\frac{1}{r} - i k\right)\frac{\exp(i k r)}{r}.$$

We make a series of approximations to this result. First, we assume that $x^2 + y^2 \ll z^2$, so that all variables except r in the exponential can be replaced by z. In addition, we assume that $\lambda \ll z$, or equivalently that $k \gg$

$1/z$, allowing us to keep only the second term in parentheses. The variable r in the exponent cannot simply be replaced by z because it is multiplied by k, which is a large number. Instead, in the exponent, we make the approximation

$$k\,r = k\,\sqrt{x^2+y^2+z^2} \approx k\,z\left(1+\frac{x^2+y^2}{2\,z^2}\right) = \frac{2\,\pi\,z}{\lambda} + \frac{\pi}{\lambda\,z}\left(x^2+y^2\right).$$

Combining all of these approximations, we find the following Fresnel approximation to the point-spread function:

$$\mathbf{h}_F(x,\,y;\,z) = \frac{1}{i\,\lambda\,z}\,e^{i\,\frac{\pi}{\lambda\,z}\left(x^2+y^2\right)},$$

where we have dropped the factor $e^{i\frac{2\pi z}{\lambda}}$ again because it corresponds to a constant phase shift across the (x, y) plane.

This form of the impulse response is much easier to work with analytically, and it is found to yield rather accurate results under most circumstances. The complex field in the diffraction field $\mathbf{U}(x,\,y;\,z)$ produced by a field $\mathbf{U}(x,\,y;\,0^+)$ in the $z = 0^+$ plane can thus be written as a convolution:

$$\mathbf{U}\,(x,\,y;\,z) = \mathbf{U}\,(x,\,y;\,0^+) \otimes \mathbf{h}_F(x,\,y;\,z) = \frac{1}{i\,\lambda\,z}\int\!\!\int_{-\infty}^{\infty} \mathbf{U}\,(\xi,\,\eta;\,0^+)\,e^{i\,\frac{\pi}{\lambda z}\left[(x-\xi)^2+(y-\eta)^2\right]}\,d\xi\,d\eta$$

$$= \frac{e^{i\,\frac{\pi}{\lambda z}\left(x^2+y^2\right)}}{i\,\lambda\,z}\int\!\!\int_{-\infty}^{\infty}\left[\mathbf{U}\,(\xi,\,\eta;\,0^+)\,e^{i\,\frac{\pi}{\lambda z}\left(\xi^2+\eta^2\right)}\right]e^{-i\,\frac{2\pi}{\lambda z}\left(\zeta\,x+\eta\,y\right)}\,d\xi\,d\eta.$$

Note that the above equation contains both a convolution solution and a solution based on a Fourier transform.

3.3 Equivalence of the Two Fresnel Approximations

We have found a Fresnel approximation in the frequency domain and a Fresnel approximation in the space domain. Are the two approximations equivalent? To answer this question, using *Mathematica®* we inverse Fourier transform the Fresnel diffraction transfer function and find the corresponding amplitude point-spread function using `ift2[ ]`, the 2D inverse Fourier transform operator defined in Chapter 1 and introduced near the beginning of this chapter in a hidden initialization cell:

```
In[ ]:= Assuming[z ∈ Reals && z ≥ 0 && λ ∈ Reals,
          ift2[Exp[-I*π*λ*z (νx^2 + νy^2)]]]
```

Out[ ]=

$$-\frac{i\,e^{\frac{i\,\pi\,(x^2+y^2)}{z\,\lambda}}}{z\,\lambda}$$

The assumptions introduced by `Assuming[ ]` are required to obtain the simplest form of the point-spread function. Remembering that $-i = 1/i$, we see that indeed the two Fresnel approximations are equivalent.

Note that while the two Fresnel approximations, one in the space domain and one in the frequency domain, have turned out to be equivalent, it is not clear exactly what restrictions are necessary for the Fresnel approximations to be accurate. This issue is a subtle one, for it is not simply a matter of the accuracy of the approximations to the square roots in both cases, but rather an issue of how accurate the approximations are under integral signs. In the case of the frequency-domain approximation, the integral comes from the inverse Fourier transform that returns us to the space domain. In the case of the space domain approximation, the integral is the convolution integral. Regions where the phase oscillation is very rapid do not contribute much to those integrals. As a consequence, accuracy is best addressed with simulations.

In the following sections we illustrate the three different methods for calculating Fresnel diffraction patterns: (1) the convolution approach, (2) the Fourier transform approach, and (3) the Fresnel transfer function approach.

3.4 A Square Aperture and the Convolution Approach

Our goal here is to illustrate the use of the convolution method for calculating diffraction patterns. Let the aperture of interest be square with width ℓ on a side and illuminated by a unit-amplitude plane wave. Thus,

$$\mathbf{U}(x, y; 0^{+}) = \mathrm{rect}(x/\ell) \times \mathrm{rect}(y/\ell).$$

Using the convolution approach, the field at distance z can be written as

$$\begin{aligned}\mathbf{U}(x, y; z) &= \frac{1}{i\lambda z}\int\int_{-\infty}^{\infty}\mathrm{rect}(\xi/\ell) \times \mathrm{rect}(\eta/\ell)\, e^{i\frac{\pi}{\lambda z}\left[(x-\xi)^2 + (y-\eta)^2\right]}\, d\xi\, d\eta \\ &= \frac{1}{\sqrt{i\lambda z}}\int_{-\ell/2}^{\ell/2} e^{i\frac{\pi}{\lambda z}\left[(x-\xi)^2\right]}\, d\xi \times \frac{1}{\sqrt{i\lambda z}}\int_{-\ell/2}^{\ell/2} e^{i\frac{\pi}{\lambda z}\left[(y-\eta)^2\right]}\, d\eta.\end{aligned}$$

The two integrals are identical, so focus attention on one of them and change the variable of integration to $\alpha = \xi/\sqrt{\lambda z}$. Then the 1D integral $\mathcal{J}_X(x)$ becomes

$$\mathcal{J}_X(x) = \frac{1}{\sqrt{i\lambda z}}\int_{-\ell/2}^{\ell/2} e^{i\frac{\pi}{\lambda z}\left[(x-\xi)^2\right]}\, d\xi = \int_{-\sqrt{N_F}}^{\sqrt{N_F}} e^{i\pi\left[\left(\alpha - 2\sqrt{N_F}\,\frac{x}{\ell}\right)^2\right]}\, d\alpha,$$

where $N_F = \frac{(\ell/2)^2}{\lambda z}$ is a dimensionless quantity known as the Fresnel number, and we have dropped a pure phase factor $1/\sqrt{i}$. We can now evaluate the integral using *Mathematica*'s **`Integrate[]`** command. Let $X = x/\ell$. Then,

```
In[•]:= Clear[nf]
        𝒥[X_, nf_] = FullSimplify[Integrate[
           Exp[I * Pi * (ξ - 2 * Sqrt[nf] * X) ^2], {ξ, -Sqrt[nf], Sqrt[nf]}]]
```

Out[•]=

$$-\frac{1}{2}(-1)^{3/4}\left(\mathrm{Erfi}\left[(-1)^{1/4}\sqrt{\mathrm{nf}}\,\sqrt{\pi}\,(1-2X)\right] + \mathrm{Erfi}\left[(-1)^{1/4}\sqrt{\mathrm{nf}}\,\sqrt{\pi}\,(1+2X)\right]\right)$$

The function **`Erfi[]`** in the solution is the imaginary error function, which is somewhat obscure but known to *Mathematica*. This function is closely related to the Fresnel integrals, also known to *Mathematica*. The integral with respect to η yields the same result, with y replacing x. To show the nature of this solution, we plot a cross-section of the 2D intensity ($I_X = |\mathcal{J}_X|^2$) vs. $X = x/\ell$ for values of N_F equal to 20, 3, 1, and 0.2, varying from close to the aperture to far from the aperture.

```
In[ ]:= nf = .2;
p1 = Plot[Abs[𝒥[𝒳, nf]]^2, {𝒳, -2, 2}, PlotRange → All,
   PlotLabel → "N_F = 0.2", Filling → Axis, AxesLabel → {"x/ℓ", "ℐ_X"}];

nf = 1;
p2 = Plot[Abs[𝒥[𝒳, nf]]^2, {𝒳, -2, 2},
   PlotRange → All, PlotLabel → "N_F = 1", Filling → Axis,
   AxesLabel → {"x/ℓ", "ℐ_X"}];

nf = 3;
p3 = Plot[Abs[𝒥[𝒳, nf]]^2, {𝒳, -2, 2},
   PlotRange → All, PlotLabel → "N_F =  7", Filling → Axis,
   AxesLabel → {"x/ℓ", "ℐ_X"}];

nf = 20;
p4 = Plot[Abs[𝒥[𝒳, nf]]^2, {𝒳, -2, 2},
   PlotRange → All, PlotLabel → "N_F =  20", Filling → Axis,
   AxesLabel → {"x/ℓ", "ℐ_X"}];

GraphicsGrid[{{p4, p3}, {p2, p1}}]

Out[ ]=
```

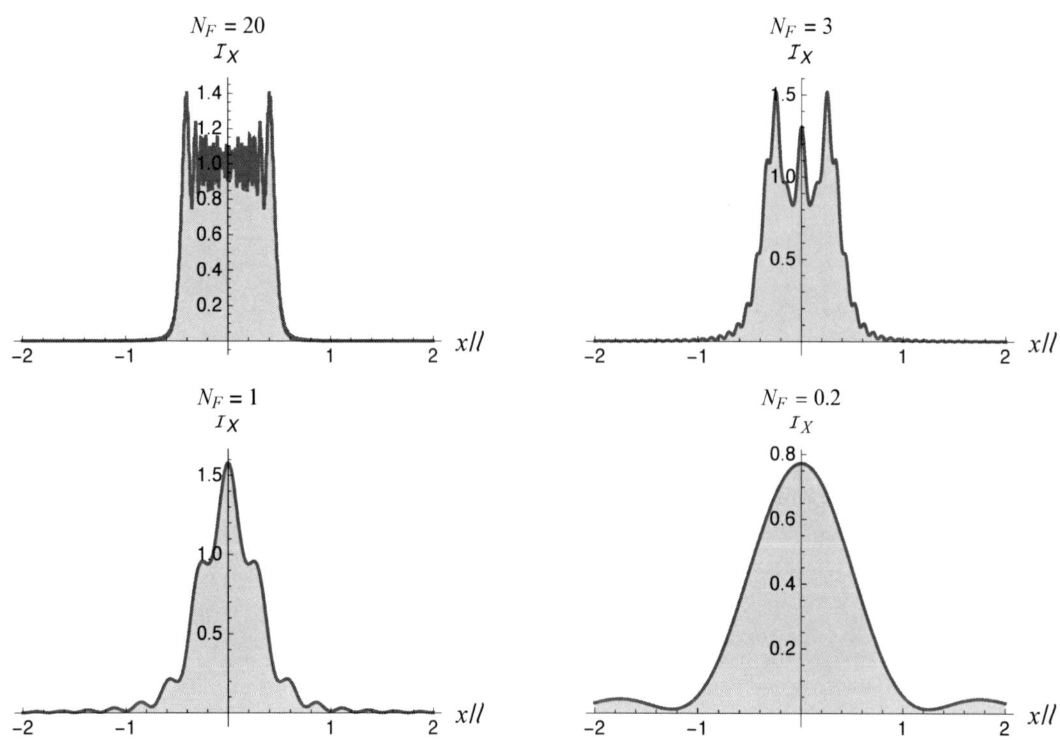

As can be seen, when the Fresnel number $N_F \gg 1$, the diffraction pattern intensity approaches the intensity distribution across the aperture itself (i.e., a rectangular distribution of width ℓ).

When $N_F \ll 1$, the pattern spreads significantly and takes on the shape of a sinc^2 function. More on this transition when we discuss Fraunhofer diffraction in Chapter 4.

3.5 A Truncated Gaussian Aperture and the Fresnel Transform Approach

Our goal here is to illustrate the use of the Fresnel transform approach to calculating diffraction patterns. For this example, we choose a $\mathbf{U}(x, y; 0^+)$ to be a truncated Gaussian function of with an amplitude value along either axis that falls to 1/2 when the argument (x or y) is (ℓ/2). This choice of width for the Gaussian makes the results more interesting than they would be if the Gaussian were much narrower than the rectangular truncation.

$$\mathbf{U}\,(x, y; 0^+) = \text{rect}[x/\ell] \times \text{rect}[y/\ell] \times e^{-\log_2(2)\left(\frac{x^2+y^2}{(\ell/2)^2}\right)}$$

It then follows that

$$\mathbf{U}(x, y; z) = \frac{1}{i\lambda z}\iint_{-\ell/2}^{\ell/2} e^{-\left(\log_e(2)\,\frac{\xi^2+\eta^2}{(\ell/2)^2}\right)}\, e^{i\frac{\pi}{\lambda z}\left(\xi^2+\eta^2\right)}\Big]\, e^{-i\frac{2\pi}{\lambda z}(\xi x+\eta y)}\, d\xi\, d\eta.$$

We can again separate this double integral into the product of two single integrals, with

$$\mathbf{U}_X(x; z) = \frac{1}{\sqrt{i\lambda z}}\int_{-\ell/2}^{\ell/2} e^{-\log_e(2)\,\frac{\xi^2}{(\ell/2)^2}}\, e^{i\frac{\pi}{\lambda z}\xi^2}\, e^{-i\frac{2\pi}{\lambda z}\xi x}\, d\xi$$

and with $\mathbf{U}_Y\,(y; z)$ given by a similar integral.

We can find a cross-section of the diffraction pattern intensity by performing the indicated 1D Fourier transform, changing the Fourier transform independent variable to $x/\lambda z$, and finding the squared modulus of the result:

```
In[ ]:= Clear[λz, ℓ, I]
        I[x_, λz_, ℓ_] = FullSimplify[
          Abs[Assuming[λz ∈ Reals && λz ≥ 0, (1 / Sqrt[I * λz]) * ft1[rect[x / ℓ] *
              Exp[-((2 * x)/ℓ)^2] * Exp[I * (Pi / λz) * x^2]] /. ν → x / λz]^2]
```

Out[ ]=

$$\frac{1}{4\,\text{Abs}\left[\pi\ell^2 + 4\,i\,\lambda z\right]}\, e^{2\pi^2\,\text{Im}\left[\frac{x^2\ell^2}{\pi\ell^2\lambda z+4\,i\,\lambda z^2}\right]}\,\pi\ell^2\,\text{Abs}\left[\text{Erfi}\left[\frac{(-1)^{1/4}\,(\pi\ell\,(-2\,x+\ell)+4\,i\,\lambda z)}{2\sqrt{(\pi\ell^2+4\,i\,\lambda z)\,\lambda z}}\right] + \text{Erfi}\left[\frac{(-1)^{1/4}\,(\pi\ell\,(2\,x+\ell)+4\,i\,\lambda z)}{2\sqrt{(\pi\ell^2+4\,i\,\lambda z)\,\lambda z}}\right]\right]^2$$

The **Erfi** function again appears. Note that *Mathematica* is treating λz as a single variable, since there is no space between the symbols to indicate that it is a product. Note also some assumptions about the variable λz that simplify the result. In this case, we have two different variables that affect the shape of the diffraction pattern, ℓ and λz. For the figures below, we fix the bounding aperture diameter ℓ to be 20 mm, and vary λz between 10^{-9} m^2 and 10^{-5} m^2.

```
In[ ]:= λz = 10^(-9);
q1=Plot[𝐼[x,λz,.002],{x,-0.002,0.002},PlotRange->All,Filling->Axis,
PlotPoints->50, AxesLabel->{"x","𝐼"},PlotLabel->"λz = 10^(-9)"];

λz = 10^(-7);
q2=Plot[𝐼[x,λz,.002],{x,-0.002,0.002},PlotRange->All,Filling->Axis,
PlotPoints->50, AxesLabel->{"x","𝐼"},PlotLabel->"λz = 10^(-7)"];

λz = 10^(-6);
q3=Plot[𝐼[x,λz,.002],{x,-.002,.002},PlotRange->All,Filling->Axis,
PlotPoints->50, AxesLabel->{"x","𝐼"},PlotLabel->"λz = 10^(-6)"];

λz = 10^(-5);
q4=Plot[𝐼[x,λz,.002],{x,-.01,.01},PlotRange->All,Filling->Axis,
PlotPoints->50, AxesLabel->{"x","𝐼"},PlotLabel->"λz = 10^(-5)"];

GraphicsGrid[{{q1,q2},{q3,q4}},ImageSize->500]
```

Out[ ]=

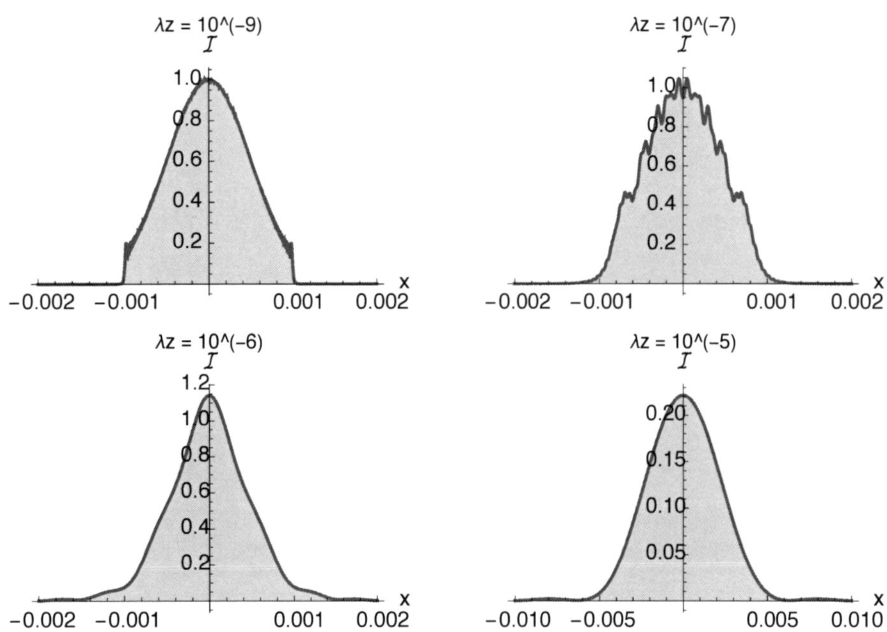

For the smallest value of λz, 10^{-9}, the observation is close enough to the aperture that we see a close replica of the truncated Gaussian function. As we move farther from the aperture (increasing λz), the diffraction pattern gradually changes, ultimately becoming a curve that resembles a blend of a sinc^2 function (the magnitude-squared Fourier transform for a rectangular aperture) and a Gaussian curve (the magnitude-squared Fourier transform for a non-truncated Gaussian aperture).

3.6 A sgn[x] × sgn[y] Function and the Fresnel Transfer Function Approach

Consider a 2D aperture that can be described analytically by amplitude transmittance $\mathbf{t}_A(x, y) = \text{sgn}(x) \times \text{sgn}(y)$. The sgn function is 1 for argument > 0 and –1 for argument < 0. The product of the two sgn functions is 1 in the upper right-hand quadrant and the lower left-hand quadrant, and –1 in the other two quadrants. This function has been chosen here to illustrate certain properties of an infinite-energy function; because it has infinite energy, it could only be approximated in practice. The plot below shows the structure of this 2D function. Our goal here is to illustrate the use of the Fresnel transfer function method for calculating diffraction patterns.

```
In[ ]:= Plot3D[sgn[x] * sgn[y], {x, -5, 5}, {y, -5, 5},
         Exclusions → "Disconinuities", AxesLabel → {x, y, "tA"}]
```

Out[ ]=

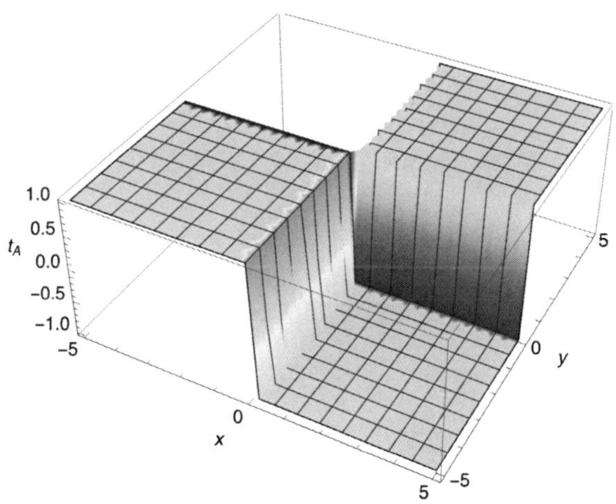

This function is separable in the (x, y) coordinate system, so it suffices to calculate a 1D transform, multiply by the 1D transfer function, inverse Fourier transform the product, and then construct a plot of the 2D intensity diffraction pattern. First calculate the 1D Fourier transform of the sgn(x) function:

```
In[ ]:= Clear[g1, g2, λz]
        g1[x_] = sgn[x];
        𝒢1[v] = ft1[g1[x]]
```

Out[ ]=

$$-\frac{i}{\pi\, v}$$

The 1D Fresnel transfer function by which we multiply this spectrum is

```
In[ ]:= ℋF[v_, λz_] = Exp[-I * Pi * λz * v^2];
```

Now plot the squared magnitude (intensity) of the inverse Fourier transform the product of the spectrum of the sgn(x) function with the 1D Fresnel transfer function:

```
In[ ]:= g2[x_, λz_] = FullSimplify[Abs[ift1[𝒢1[v] * ℋF[v, λz]]]^2]
```

Out[ ]=

$$\text{Abs}\left[\text{Erfi}\left[\frac{(-1)^{1/4}\sqrt{\pi}\,x}{\sqrt{\lambda z}}\right]\right]^2$$

In[ ]:=
```
λz = 10^{-7};
w1 = Plot[g2[x, λz], {x, -0.005, 0.005}, PlotLabel → "λz = 10^-7",
   PlotRange → All, Filling → Axis, ImagePadding → 10];

λz = 10^{-6};
w2 = Plot[g2[x, λz], {x, -0.013, 0.013}, PlotLabel → "λz = 10^-6",
   PlotRange → All, Filling → Axis, ImagePadding → 10];

λz = 10^(-5);
w3 = Plot[g2[x, λz], {x, -0.05, 0.05}, PlotLabel → "λz = 10^-5",
   PlotRange → All, Filling → Axis, ImagePadding → 10];

λz = 10^(-4);
w4 = Plot[g2[x, λz], {x, -0.15, 0.15}, PlotLabel → "λz = 10^-4",
   PlotRange → All, Filling → Axis, ImagePadding → 10];

GraphicsGrid[{{w1, w2}, {w3, w4}}, ImageSize → 500]
```

Out[ ]=

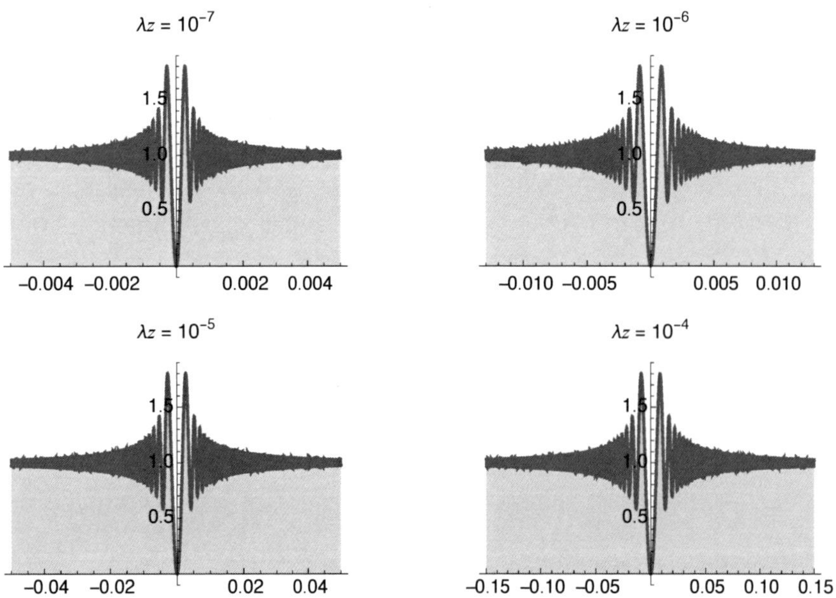

Intensity is plotted vertically, and position in the pattern is plotted horizontally. As λz increases, the pattern expands but the shape and height of the pattern remain unchanged. The lack of a change of height is related to the fact that the function we are transforming has infinite energy. To relate the results to real-world dimensions, note that if $\lambda = 10^{-6}$ m, then the case $\lambda z = 10^{-7}$ corresponds to a viewing distance of 10 cm. For this wavelength, then, the viewing distances for the remaining three curves are 1 m, 10 m, and 100 m.

Note that the diffraction patterns all have lines of zero value for $x = 0$ where the original function sgn(x) has value zero. This is a consequence of the fact that a real and odd function has an imaginary and odd spectrum. The squared-magnitude operation to show intensity changes the odd and imaginary amplitude distribution to a real and even intensity distribution.

The examples above have all been for apertures that are separable in the (x, y) coordinate system. We now turn to the case of circularly symmetric apertures.

3.7 The Zero-Order Hankel Transform

There are at least two different methods for calculating the diffraction patterns of circularly symmetric aperture functions (i.e., functions of only radius r; see Chapter 6 of Ref. [1]). The first of these methods is the zero-order Hankel transform.

The 2D Fourier transform can be reduced in the circularly symmetric case to the following equation:

$$\mathcal{G}(\rho) = 2\pi \int_0^\infty r\,\mathbf{g}(r)\,J_0(2\pi r\rho)\,dr,$$

where $\rho = \sqrt{\nu_x^2 + \nu_y^2}$ represents the radius in the Fourier transform plane and $J_0(\,)$ is a Bessel function of the first kind, zero order. Since the quadratic-phase exponential in the Fresnel diffraction equation is also circularly symmetric, we can write the following equation for the field amplitude $\mathbf{U}(r; z)$ in the diffraction pattern of a circularly symmetric aperture function $\mathrm{U}(r; 0)$:

$$\mathbf{U}(r; z) = \frac{2\pi}{\lambda z} e^{i\pi\frac{r^2}{\lambda z}} \int_0^\infty \xi\,\mathbf{U}(\xi; 0)\,e^{i\pi\frac{\xi^2}{\lambda z}}\,J_0\!\left(\frac{2\pi r\xi}{\lambda z}\right) d\xi.$$

This integral relationship is a form of the Hankel transform of zero order. Turning to a more specific integral, consider a uniform circular aperture with diameter ℓ as the diffracting structure. The integral becomes

$$\mathbf{U}(r; z) = \frac{2\pi}{\lambda z} e^{i\pi\frac{r^2}{\lambda z}} \int_0^{\ell/2} \xi\,e^{i\pi\frac{\xi^2}{\lambda z}}\,J_0\!\left(\frac{2\pi r\xi}{\lambda z}\right) d\xi.$$

As in the case with the square aperture, make a change of variables $\alpha = \xi/\sqrt{\lambda z}$, and define the Fresnel number $N_F = (\ell/2)^2/\lambda z$, yielding

$$\mathbf{U}(r; z) = 2\pi\,e^{i\frac{\pi r^2}{\lambda z}} \int_0^{\sqrt{N_F}} \alpha\,e^{i\pi\alpha^2}\,J_0\!\left(\sqrt{N_F}\,\frac{2\pi r\alpha}{\ell/2}\right) d\alpha.$$

Dropping the quadratic phase factor that precedes the integral since it does not affect the intensity in the diffraction pattern, and defining a normalized radius $r_0 = r/\ell$, the complex field expressed in *Mathematica* language is

```
In[•]:= u[r0_, nf_] := 2*Pi*Integrate[α*Exp[I*Pi*α^2] *
          BesselJ[0, Sqrt[nf] 2*Pi*(2*r0)*α], {α, 0, Sqrt[nf]}]
```

The "delayed equal" sign **:=** in the *Mathematica* equation means that the integral will not be evaluated until it is called in a subsequent command. *Mathematica* cannot find a symbolic answer to this integral, but will use numerical integration if it is called in a **Plot[]** command. Therefore, we plot the intensity diffraction pattern for vs. $r_0 = r/\ell$ for various values of the Fresnel number N_F. The plotting computations are sufficiently lengthy that we have carried them out in a separate notebook and pasted the results below:

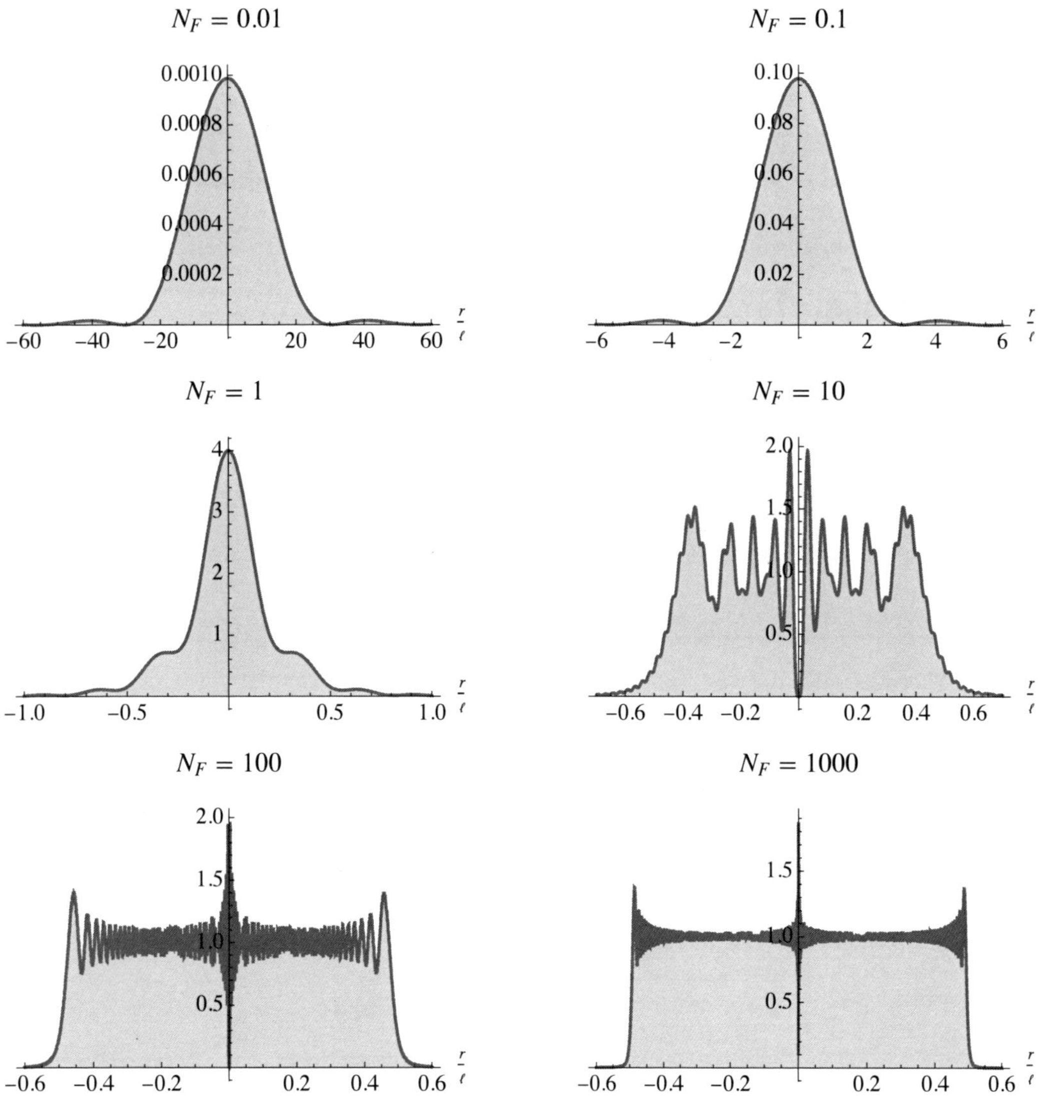

Note that the larger the Fresnel number, the closer the diffraction pattern resembles the diffracting aperture, while the smaller the Fresnel number, the more the diffraction pattern spreads and smooths.

The results for $N_F = 10$, 100, and 1000 all have a zero of intensity on axis. This is a result of the contributions of opposite signs from various circular rings on the aperture interfering with one another to create zero intensity, a phenomenon that occurs only for large N_F.

3.8 The Projection-Transform Method

An alternative method for finding the diffraction patterns of circularly symmetric apertures is a method called the projection-transform method. This method is based on the projection-slice theorem of Fourier analysis, which is discussed in Ref. [1], Section 6.3. Summarizing the theorem for this application, the 1D Fourier transform of any projection through a circularly symmetric function is equal to a central slice through the 2D spectrum of that function. Mathematically, if $\mathbf{g}(r)$ is the amplitude transmittance of a circularly symmetric aperture, then its projection onto the x axis (although we could project onto any axis) is given by

$$\mathbf{p}(x) = \int_{-\infty}^{\infty} \mathbf{g}(x, y)\, dy.$$

Then the projection-slice theorem states that the 1D Fourier transform $\mathbf{P}(\nu)$ of $\mathbf{p}(x)$ is equal to a slice $\mathbf{G}(\nu_x, 0)$ along the ν_x axis of the 2D Fourier transform $\mathbf{G}(\nu_x, \nu_y)$. Thus, we can in principle find the radial profile of the Fourier transform of any circularly symmetric function by this method.

The projection-transform method is entirely equivalent to performing a 2D transform with the frequency variable ν_y set equal to zero. In that case, the integral with respect to y is simply a projection onto the x axis. Thus, to find the complex amplitude of the Fresnel diffraction pattern $\mathbf{g}(r; z)$ of the aperture function $\mathbf{g}(r; 0)$, we can perform the operation

$$\mathbf{g}(r; z) = \frac{e^{i \frac{\pi r^2}{\lambda z}}}{\lambda z} \int_{-\infty}^{\infty} \left[\int_{-\infty}^{\infty} \mathbf{g}\left(\sqrt{\xi^2 + \eta^2}\ ; 0\right) e^{i \frac{\pi}{\lambda z} \eta^2}\, d\eta \right] e^{-i \frac{2\pi}{\lambda z} x \xi}\, d\xi,$$

where $r = \sqrt{x^2 + y^2}$.

This approach works well for finding the Fourier transform of some circularly symmetric functions, but not very well for finding the Fresnel diffraction patterns of most symbolic circularly symmetric functions. One case in which it does work well is when finding the radial profile of the Fresnel diffraction pattern of an aperture with a Gaussian radial profile extending from 0 to ∞. While no aperture radial profile can extend to infinity, such a profile is an approximation to the radial profile of the lowest-order mode of a gas laser. We illustrate the method as follows. First, define the aperture function of interest and the integrand of the Fresnel diffraction integral:

```
In[•]:= Clear[h, g, λz, f]
        r = Sqrt[x^2 + y^2];
        h[r_] = gaus[r];
        g[r_] = h[r] * (1 / λz) * Exp[I * Pi * r^2 / λz];
```

Note that, throughout this calculation, the combination of λ and z, i.e., λz, is treated by *Mathematica* as a single symbol, since there is no space or multiplication sign between the two symbols. Now find the projection $\mathbf{p}(x)$ of $\mathbf{g}(r)$ onto the x axis:

```
In[•]:= p[x_] = Assuming[λz ∈ Reals && λz > 0,
          Integrate[g[r], {y, -Infinity, Infinity}]]
Out[•]=
```

$$\frac{(-1)^{1/4}\, e^{-\frac{\pi x^2 (-i + \lambda z)}{\lambda z}}}{\sqrt{\lambda z}}$$

Now find the 1D Fourier transform of the projection, with frequency ν scaled by $1/\lambda z$:

```
In[•]:= f[ξ_] = ft1[p[x]] /. ν → ξ / λz
Out[•]=
```

$$\frac{(-1)^{1/4}\, e^{-\frac{\pi \xi^2}{\lambda z (-i + \lambda z)}}}{\sqrt{\lambda z}\, \sqrt{\frac{-i + \lambda z}{\lambda z}}}$$

The symbols `/. ν→ξ/λz` are an assignment command, resulting in the replacement of ν by $\xi/\lambda z$. The `ComplexExpand[]` command simplifies the result. Now find and plot the squared absolute value of this result:

In[ ]:=
```
Clear[λz]
field[ξ_, λz_] = ComplexExpand[Abs[f[ξ]]^2]
```

Out[ ]=

$$\frac{e^{-\frac{2\pi\xi^2}{1+\lambda z^2}}}{\sqrt{1+\lambda z^2}}$$

In[ ]:=
```
d1 = Plot[field[ξ, 0], {ξ, 0, 10},
   PlotRange → All, AxesLabel → {x, "Intensity"},
   PlotLabel → "λz = 0", PlotRange → All, Filling → Axis];

d2 =
  Plot[field[ξ, 3], {ξ, 0, 10}, PlotRange → All, PlotLabel → "λz = 3",
   AxesLabel → {x, "Intensity"}, PlotRange → All, Filling → Axis];

d3 = Plot[field[ξ, 6], {ξ, 0, 10}, PlotRange → All,
   AxesLabel → {x, "Intensity"}, PlotLabel → "λz = 6",
   PlotRange → All, Filling → Axis, ImagePadding → 20];

d4 = Plot[field[ξ, 9], {ξ, 0, 10},
   PlotRange → All, AxesLabel → {x, "Intensity"},
   PlotLabel → "λz = 9", PlotRange → All, Filling → Axis];

GraphicsGrid[{{d1, d2}, {d3, d4}}, ImageSize → 500]
```

Out[ ]=

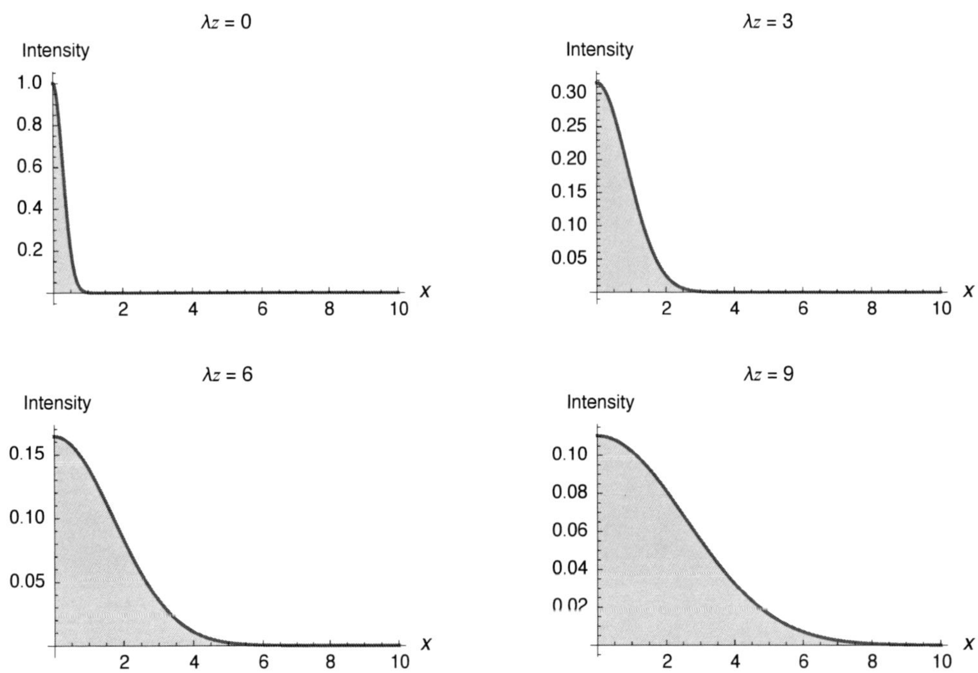

Note that when $\lambda z = 0$, the radial profile of the diffraction pattern is the squared magnitude of the radial profile of the aperture function. As λz increases, the radial profile of the diffraction pattern broadens and drops in height, the latter to conserve energy. The shape of the diffraction pattern remains Gaussian for all λz.

The projection-transform method is quite limited for symbolically defined apertures. I have found no other symbolic apertures for which it works other than the Gaussian aperture, but some may exist. However, this method will be very useful when calculating Fresnel diffraction patterns of discrete (sampled) aperture functions; see the discussion in Chapter 6.

4. Fraunhofer Diffraction Patterns

As the Fresnel number becomes less than unity (or equivalently, as z becomes sufficiently large), the quadratic-phase exponential under the integral sign in the Fresnel diffraction equation approaches unity, leaving a simple scaled Fourier transform of the diffracting aperture amplitude transmittance as the diffracted amplitude pattern. Thus, in the Fraunhofer regime,

$$\mathbf{g}(x, y; z) = \frac{e^{i \frac{\pi}{\lambda z}(x^2+y^2)}}{\lambda z} \int\int_{-\infty}^{\infty} \mathbf{g}(\xi, \eta; 0)\, e^{-i \frac{2\pi}{\lambda z}(x\xi + y\eta)}\, d\xi\, d\eta.$$

The quadratic-phase exponential that precedes the integral can be neglected if only the intensity distribution in the diffraction pattern is of interest. A more extensive discussion of this approximation can be found in Ref. [4].

We now give some examples of diffraction-pattern calculations.

4.1 Square Aperture

Let the dimensions of the square aperture be $\ell \times \ell$. Then the corresponding amplitude transmittance of the aperture is

$$\mathbf{t}_A(\xi, \eta) = \text{rect}(\xi/\ell) \times \text{rect}(\eta/\ell).$$

First find the Fourier transform of the aperture function $\mathbf{t}_A$.

```
In[ ]:= Assuming[ℓ ∈ Reals && ℓ > 0, (1 / λz) ^2 * ft2[rect[ξ / ℓ] * rect[η / ℓ]] ^2]
```

$$\frac{\ell^4\, \text{Sinc}[\pi \ell\, \nu x]^2\, \text{Sinc}[\pi \ell\, \nu y]^2}{\lambda z^2}$$

Remember that the *Mathematica*® `Sinc[x]` function is equivalent to the user-defined `sinc[x/π]` function.

Find the maximum value of the intensity and then normalize the full result by this maximum before plotting:

```
In[ ]:= max = Limit[%, {νx → 0, νy → 0}]
```

$$\frac{\ell^4}{\lambda z^2}$$

Now plot the normalized result, showing the 2D intensity pattern and a cross-section of that pattern, side by side.

```
In[ ]:= p1 = Plot3D[Sinc[Pi * νx] ^2 * Sinc[Pi * νy] ^2, {νx, -2, 2}, {νy, -2, 2},
          PlotRange → {0, 1}, AxesLabel → {ℓ x / λz , ℓ y / λz, I}];
        p2 = Plot[Sinc[Pi * νx] ^2, {νx, -2, 2}, AxesLabel → {ℓ x / λz, I} ];
        GraphicsRow[{p1, p2}, ImageSize → 500]
```

Out[ ]=

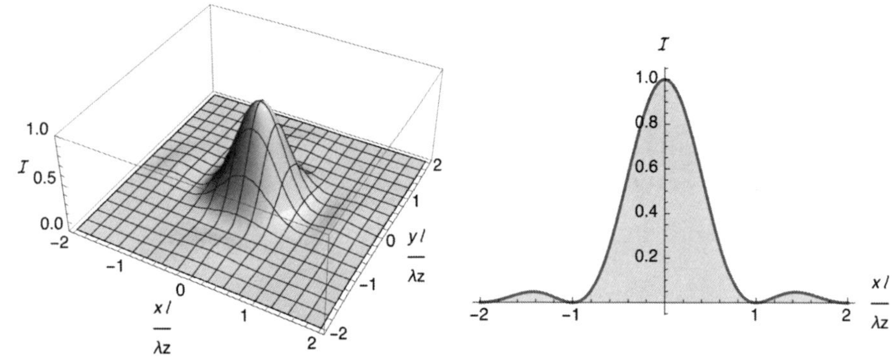

As seen from the cross-section of intensity, the zeros of intensity fall at the non-zero integer values of $\frac{x\ell}{\lambda z}$.

4.2 Circular Aperture

Consider now a circular diffracting aperture with diameter ℓ. Then the corresponding amplitude transmittance of the aperture is

$$\mathbf{t}_A(\xi, \eta) = \text{circ}\left(2\sqrt{\xi^2+\eta^2}\,/\,\ell\right).$$

We perform a 2D Fourier transform of the aperture amplitude transmittance. To obtain a simple result, we must make certain assumptions about the real-valued and non-negative properties of some of the parameters. We take the square of the Fourier transform to obtain intensity.

In[ ]:=
```
unnormalized[vx_, vy_] = FunctionExpand[Assuming[
    ℓ ∈ Reals && ℓ > 0 && λz ∈ Reals && λz ≥ 0 && Sqrt[vx^2 + vy^2] > 0,
    (1 / λz)^2 * ft2[circ[2 * Sqrt[ξ^2 + η^2] / ℓ]]^2]]
```

$$\frac{\ell^2\,\text{BesselJ}\left[1, \pi\,\ell\,\sqrt{\text{vx}^2+\text{vy}^2}\right]^2}{4\,\lambda\text{z}^2\,\left(\text{vx}^2+\text{vy}^2\right)}$$

To normalize this result to unity at the origin, we must find its value at the origin:

In[ ]:=
```
origin = Limit[unnormalized[vx, vy], {vx → 0, vy → 0}]
```

$$\frac{\pi^2\,\ell^4}{16\,\lambda\text{z}^2}$$

Now normalize by the value of the intensity at the origin:

In[ ]:=
```
normalized = Assuming[ℓ ∈ Reals && ℓ > 0, Simplify[
    (unnormalized[vx, vy] / origin) ]] /. {vx → x / λz, vy → y / λz}
```

$$\frac{4\,\text{BesselJ}\left[1, \pi\,\ell\,\sqrt{\frac{\text{x}^2}{\lambda\text{z}^2}+\frac{\text{y}^2}{\lambda\text{z}^2}}\right]^2}{\pi^2\,\ell^2\,\left(\frac{\text{x}^2}{\lambda\text{z}^2}+\frac{\text{y}^2}{\lambda\text{z}^2}\right)}$$

In finding the function **normalized** in its simplest form, we had to introduce a new command, **FunctionExpand[]**, which reduces some complicated functions to simpler forms.

To simplify this expression, define two new variables $\alpha = \ell\, \nu_x = t\, x/\lambda z$ and $\beta = \ell\, \nu_y = t\, y/\lambda z$.

```
In[•]:= normalized1 = Assuming[ℓ ∈ Reals && ℓ > 0,
          Simplify[normalized /. {x → λz α / ℓ, y → λz β / ℓ}]]
```

$$\frac{4\,\text{BesselJ}\left[1,\ \pi\sqrt{\alpha^2+\beta^2}\,\right]^2}{\pi^2\left(\alpha^2+\beta^2\right)}$$

Make a 2D plot of the intensity pattern, as well as a cross-sectional slice of that pattern through the origin.

```
In[•]:= p1 = Plot3D[normalized1, {α, -2, 2}, {β, -2, 2},
          AxesLabel → {ℓ x / (λz ), ℓ y / (λz), I}, PlotRange → All];
        p2 = Plot[normalized1 /. β → 0, {α, -2, 2}, AxesLabel → {ℓ x / (λz) , I}];
        GraphicsRow[{p1, p2}, ImageSize → 500]
Out[•]=
```

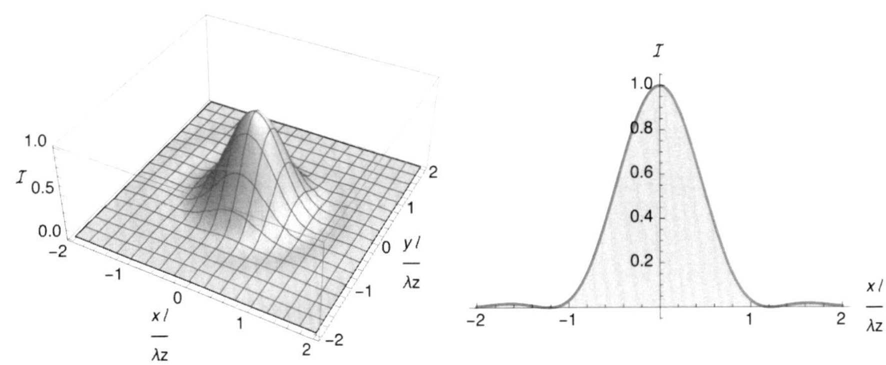

Note that the circularly symmetric diffraction pattern falls to zero at radius $\pm 1.22\,\lambda z/\ell$.

We have assumed a circular aperture in finding these limits. For most optical problems, the aperture is in practice bounded by a circular aperture. However, keep in mind that some apertures, while bounded by a circle, may have the majority of their significant amplitude transmittance restricted to a smaller diameter. For example, a truncated Gaussian aperture with a $1/e$ width that is much smaller than the diameter of the bounding circle will have an effective Fresnel number that is smaller than that of the circular aperture, and may transition from Fresnel to Fraunhofer diffraction at a Fresnel number that is larger than that appropriate for the clear circle.

4.3 When is the Fraunhofer Approximation Accurate?

It is of some interest to identify a rough boundary between the Fresnel diffraction domain and the Fraunhofer diffraction domain. There are several approaches to doing so, all of which yield roughly the same result. As a first approach, we may require the exponent of the quadratic-phase exponential under the integral sign to be small enough so as not to change the phase by very much. A suitable requirement might be that at maximum, the change of phase in the exponent is less than $\pi/4$ radians. For a circular aperture of diameter ℓ,

$$\frac{\pi}{\lambda z}\left(\xi^2+\eta^2\right)_{\max} = \frac{\pi}{\lambda z}(\ell/2)^2 < \pi/4 \qquad \text{or} \qquad N_F < 1/4 = 0.25.$$

Note, however, that it is not strictly necessary that the phase of the quadratic-phase exponential be less than some number such as $\pi/4$. Rather, it is necessary that the quadratic-phase exponential have no effect on the Fresnel-diffraction integral. This can happen due to the rapid oscillations of that exponential under the integral sign, so other methods of estimating the required Fresnel number are of interest.

Another criterion often cited is the so-called "antenna designer's formula," which states that a sufficient condition to assure that Fraunhofer diffraction applies is

$$z > \frac{2\ell^2}{\lambda} \quad \text{or} \quad N_F < 1/8 = 0.125.$$

This condition is slightly more stringent that the previous one.

A final approach considers the form of the intensity distribution of the Fresnel diffraction pattern of a circular aperture as the Fresnel number transitions from 0.5 to 0.25; two plots for these Fresnel numbers are shown below, calculated using numerical integration to speed up the results.

```
In[ ]:= Clear[α, r0]
        u[r0_, nf_] := 2 * Pi * NIntegrate[α * Exp[I * Pi * α^2] *
            BesselJ[0, Sqrt[nf] 2 * Pi * (2 * r0) * α], {α, 0, Sqrt[nf]}]

In[ ]:= Plot[{Abs[u[r0, 0.5] / u[0, 0.5]]^2, Abs[u[r0, 0.25] / u[0, 0.25]]^2},
         {r0, -2, 2}, Filling → None, PlotStyle → {Blue, , Red},
         PlotLegends → {"N_F=0.5", "N_F=0.25"}, AxesLabel → {ℓ r / λ z, I}]
Out[ ]=
```

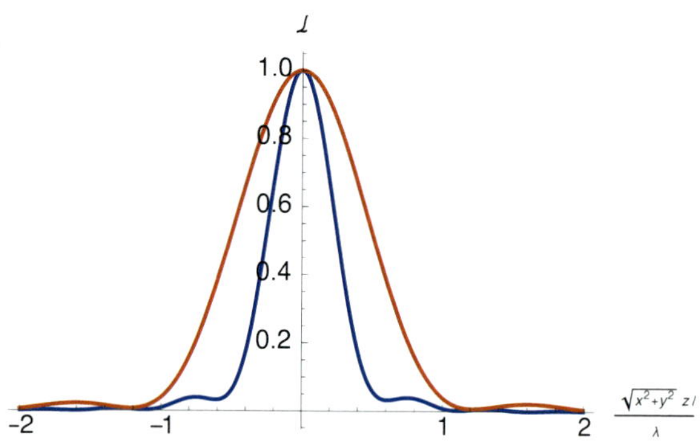

We see from these plots that the Fresnel intensity pattern for $N_F = 0.5$ (blue) has not yet attained its $(J_1(r)/r)^2$ shape, but when the Fresnel number is reduced to 0.25 (red), there is a close resemblance to that shape. In particular, the first zero of the diffraction pattern intensity falls at a radius r of approximately $1.22\ \lambda z/\ell$. We conclude from the curves above that for $N_F < 1/4 = 0.25$, the Fraunhofer conditions can reasonably be assumed to hold.

5. Diffraction by an Arbitrary Linear Optical System

We first introduce the concept of ray-transfer matrices (also known as ABCD matrices) to describe the geometrical optics mapping introduced by passage of light through an arbitrary linear optical system. Then we present a diffraction formula that is determined by the system's ray transfer matrix describing the mapping from input fields to output fields. This approach to diffraction calculations is restricted to optical systems that do not contain internal limiting apertures. However, if the system does contain a finite aperture, this formalism can be used up to the aperture, the aperture limitation can be introduced, and the same formalism can then be used to pass from the aperture to the output plane. For an alternative discussion of these topics without *Mathematica®* code, see Ref. [4], Appendix B.

5.1 Ray-Transfer (or ABCD) Matrices

The theory of ray-transfer matrices starts with two assumptions: (1) paraxial optics holds, so sines and tangents of angles can be replaced by the angles themselves while cosines of angles can be replaced by unity; and (2) only rays confined to a meridional plane [in this case, the (y, z) plane] are treated.

A geometrical ray is assumed to enter a linear optical system with vertical coordinate y_1 and *reduced* angle $\hat{\theta}_1$, where a reduced angle is an actual angle with respect to the z axis divided by the local refractive index at the input. That same ray exits the optical system with vertical coordinate y_2 and reduced angle $\hat{\theta}_2$, where again the reduced angle is the actual angle divided by the local refractive index at the output. The y_2-coordinate of the output is determined by the y_1-coordinate of the input, the reduced angle $\hat{\theta}_1$ at the input, and the properties of the optical system through which the ray passes. Likewise, the reduced angle at the output is determined by both input parameters and the characteristics of the optical system through which the ray has passed. The transformation of the input parameters to the output parameters is represented by a ray-transfer matrix $\mathcal{M}$ (also called an ABCD matrix) that maps the input vector $\begin{pmatrix} y_1 \\ \hat{\theta}_1 \end{pmatrix}$ into an output vector $\begin{pmatrix} y_2 \\ \hat{\theta}_2 \end{pmatrix}$. The form of the matrix mapping can be written as

$$\begin{pmatrix} y_2 \\ \hat{\theta}_2 \end{pmatrix} = \begin{pmatrix} \mathcal{A} & \mathcal{B} \\ \mathcal{C} & \mathcal{D} \end{pmatrix} \begin{pmatrix} y_1 \\ \hat{\theta}_1 \end{pmatrix},$$

where $\mathcal{A}$ maps y_1 into y_2, $\mathcal{B}$ maps $\hat{\theta}_1$ into y_2, etc. A few examples of ABCD matrices will be instructive.

Propagation of a Ray Through Free Space

Let the optical system consist of a region of free space of width z with refractive index n. As the ray passes through this region, the reduced angle stays constant, but the exit position of the ray is determined by the reduced angle of incidence and the width z of the region divided by the refractive index. We have $y_2 = y_1 + \frac{z}{n}\,\hat{\theta}_1$ and $\hat{\theta}_2 = \hat{\theta}_1$. Thus, the ray-transfer matrix is $\begin{pmatrix} 1 & z/n \\ 0 & 1 \end{pmatrix}$.

We represent this matrix in *Mathematica* language as follows:

```
In[∘]:= Clear[z, n]
        m1[z_, n_] = {{1, z/n}, {0, 1}}; MatrixForm[m1[z, n]]
```

$$\begin{pmatrix} 1 & \frac{z}{n} \\ 0 & 1 \end{pmatrix}$$

Propagation of a Ray Through a Planar Interface

With similar reasoning, we find that propagation of a ray through a planar interface between two infinite media with different refractive indices is governed by the ray-transfer matrix:

$$\mathcal{M} = \begin{pmatrix} 1 & 0 \\ 0 & 1 \end{pmatrix}.$$

Here the input and output of the optical system are the planes on the left face of the interface and on the right face of the interface, respectively. Note that the reduced angle is not changed ($\mathcal{D} = 1$).

Propagation of a Ray Through a Spherical Interface

Consider now a spherical interface between two regions of different refractive indices n_1 and n_2. The radius of curvature of the sphere is represented by $\mathcal{R}$, which we take to be positive for a convex surface and negative for a concave surface. A ray striking this surface at position y exits the surface at the same y. Thus, $\mathcal{A} = 1$. At a point y on the interface, the reduced transmitted angle is not the same as the reduced incidence angle, due to the tilt of the spherical surface at that coordinate. The normal to the surface is not parallel to the optical axis, but rather is tilted with respect to that axis by angle $\Theta = \arcsin\frac{y}{\mathcal{R}} \approx \frac{y}{\mathcal{R}}$. Snell's law of geometrical optics states that at an interface between a region of refractive index n_1 and a region of n_2, the incident ray angle θ_1 and the transmitted ray angle θ_2, both measured with respect to the optical axis, must obey

$$\frac{\sin\theta_2}{\sin\theta_1} \approx \frac{\theta_2}{\theta_1} = \frac{n_1}{n_2}.$$

From this result, we deduce that

$$n_1\,\theta_1 + n_1\,\frac{y}{\mathcal{R}} = n_2\,\theta_2 + n_2\,\frac{y}{\mathcal{R}}.$$

It follows that, using reduced angles,

$$\hat{\theta}_2 = \hat{\theta}_1 + \frac{n_1 - n_2}{\mathcal{R}}\,y,$$

and the ray-transfer matrix must be

```
In[∘]:= 𝒜 = 1; ℬ = 0; 𝒞 = (n1 - n2) / ℛ; 𝒟 = 1;
        m2 = {{𝒜, ℬ}, {𝒞, 𝒟}}; MatrixForm[m2]
```

$$\begin{pmatrix} 1 & 0 \\ \frac{n1-n2}{\mathcal{R}} & 1 \end{pmatrix}$$

Passage Through a Thin Lens

A beauty of ray-transfer analysis is that when optical propagation or passage through interfaces are cascaded, the resulting ray-transfer matrix is the product of the component ray-transfer matrices, but applied from right to left. Thus, for a thin lens with one convex surface between a medium with index n_1 and a medium with index n_2, followed immediately by a concave surface between n_2 and n_1, the overall transfer matrix will be

```
In[]:= m3 = {{1, 0}, {(n1 - n2) / ℛ2, 1}}.{{1, 0}, {(n2 - n1) / ℛ1, 1}};
       MatrixForm[m3]
```

$$\begin{pmatrix} 1 & 0 \\ \frac{-n1+n2}{\mathcal{R}1} + \frac{n1-n2}{\mathcal{R}2} & 1 \end{pmatrix}$$

This expression can be simplified by defining the focal length of the lens as

$$\frac{1}{f} = \frac{n_2 - n_1}{n_1}\left(\frac{1}{\mathcal{R}_1} - \frac{1}{\mathcal{R}_2}\right),$$

yielding a ray-transfer matrix:

```
In[]:= Clear[m3, f]
       m3 = {{1, 0}, {-n1 / f, 1}}; MatrixForm[m3]
```

$$\begin{pmatrix} 1 & 0 \\ -\frac{n1}{f} & 1 \end{pmatrix}$$

We now have sufficient information to find the ray-transfer matrix of cascades of sections of free space and thin lenses.

5.2 Wave Analysis of Complex Optical Systems Using Ray-Transfer Matrices

While ray-transfer matrices are derived using geometrical optics, they can play a very useful role in a full scalar wave theory of the passage of light through optical systems. The relation we are about to present requires a more detailed and lengthy proof than we wish to present here, but we refer the reader to Ref. [6]. The result of interest can be written as the equation below:

$$\mathbf{U}(y; z_0) = \frac{e^{i\frac{2\pi L_0}{\lambda_0}}}{\sqrt{i\mathcal{B}\lambda}} \int_{-\infty}^{\infty} \mathbf{U}(\xi; 0) \exp\left[\frac{i\pi}{\mathcal{B}\lambda}\left(\mathcal{A}\xi^2 - 2y\xi + \mathcal{D}y^2\right)\right] d\xi,$$

where $\mathcal{A}$, $\mathcal{B}$, and $\mathcal{D}$ are the corresponding elements of the ray-transfer matrix, λ_0 is the vacuum wavelength of light, and L_0 is the total optical distance (i.e., the refractive index times the physical distance) traveled from input to output down the optical axis. Thus, if we know the total ray-transfer matrix of an optical system, possibly containing many lenses and distances of free-space propagation, then the full amplitude and phase of the light at the output, including the effects of diffraction, can be obtained by the relationship given. However, the optical system should not contain any internal apertures, for the effects of such apertures are not included in this formalism.

This relationship is obviously one-dimensional (i.e., it only finds a y distribution of the complex amplitude). If the optical system is separable in x and y, then a similar analysis can be performed in the x direction. We turn now to some examples of the use of the general diffraction equation.

Generalized Transfer Function of Propagation Through a Complex Optical System

By analogy with the transfer function approach to Fresnel diffraction, there is a transfer function approach that incorporates the element of the ABCD matrix. We can find that transfer function by letting $\mathbf{U}(\xi; 0)$ in the equation above be a Dirac delta function $\delta(\xi)$, and thus we first find the impulse response of the system in terms of the elements of the ABCD matrix. Once the impulse response has been found, a Fourier transform will yield the corresponding transfer function.

We begin by finding the impulse response using *Mathematica*:

```
In[*]:= Clear[𝒜, ℬ, 𝒞, 𝒟]
        h = FullSimplify[Assuming[λℬ ∈ Reals && y ∈ Reals,
           1 / Sqrt[I * λ ℬ] * Exp[I * Pi / (λ ℬ) * (𝒟 * y^2) * Integrate[
              DiracDelta[ξ] * Exp[I * (Pi / (λ ℬ)) * (𝒜 * ξ^2 - 2 * y * ξ)],
              {ξ, -Infinity, Infinity}]]]]
```

$$\frac{e^{\frac{i \pi y^2 \mathcal{D}}{\mathcal{B} \lambda}}}{\sqrt{i \mathcal{B} \lambda}}$$

Note that the element $\mathcal{D}$ can be positive or negative, depending on the problem. The impulse response is a quadratic-phase exponential in y times a complex pre-multiplier. Now Fourier transform the impulse response to find the transfer function:

```
In[*]:= ℋ[v_] = ft1[h]
```

$$\frac{e^{-\frac{i \pi \mathcal{B} \lambda v^2}{\mathcal{D}}}}{\sqrt{-\frac{i \mathcal{D}}{\mathcal{B} \lambda}} \sqrt{i \mathcal{B} \lambda}}$$

We can simplify this result with cancellations, yielding a transfer function:

$$\mathcal{H}(\nu) = \frac{1}{\sqrt{\mathcal{D}}} e^{-i \pi \mathcal{B} \lambda \left(\frac{\nu}{\sqrt{\mathcal{D}}}\right)^2}.$$

The approach using the transfer function has advantages in some problems.

Operator for Propagation Through a Complex Optical System

For a given ABCD matrix, it is possible to define an operator that will calculate the result of an input field propagating through the system to form an output complex field. We define that operator through the following *Mathematica* code:

```
In[*]:= propagate[func_] := 1 / Sqrt[I * λ * ℬ] * Exp[I * (Pi / Sqrt[λ * z]) * 𝒟 * y^2]
          Integrate[func * Exp[I * (Pi / (λ * ℬ)) * (𝒜 * ξ^2 - 2 * y * ξ)],
           {ξ, -Infinity, Infinity}]
```

Here *func* represents the field that is presented at the input to the system, and the result of applying the operator is the field found at the output.

We next apply this operator to several different optical geometries.

Diffraction of a Space-Limited Spherical Wave Converging Towards the Observation Plane

In this example, we assume that a paraxial approximation to a converging spherical wave propagates down the optical axis, with its point of convergence lying on the z axis in a plane at distance z from the plane in which we specify the spherical wave. The complex amplitude of the spherical wave in the plane $z = 0$ is described by a function **g**(ξ). We also assume that the refractive index in the propagation medium is unity. A paraxial representation of the space-limited converging wave in the plane $z = 0$ is given by

$$\mathbf{U}(\xi; 0) = \exp\left(-\frac{i\pi}{\lambda z}\xi^2\right)\mathbf{g}(\xi).$$

This wave then propagates over distance z. The matrix for propagation through free space over distance z with $n = 1$ is

```
In[]:= m1[z_, 1] = {{1, z}, {0, 1}}; MatrixForm[m1[z, 1]]
```

$$\begin{pmatrix} 1 & z \\ 0 & 1 \end{pmatrix}$$

It follows that

```
In[]:= 𝒜 = 1; ℬ = z; 𝒞 = 0; 𝒟 = 1;
```

We now apply the propagation operator to the field above:

```
In[]:= FullSimplify[propagate[Exp[-I π ξ^2 / (λ z)] * g[ξ]]]
```

$$\frac{e^{\frac{i\pi y^2}{\sqrt{z\lambda}}}\int_{-\infty}^{\infty} e^{-\frac{2 i \pi y \xi}{z\lambda}}\, g[\xi]\, d\xi}{\sqrt{i\, z\, \lambda}}$$

We see that the diffraction integral reduces to a scaled Fourier transform of the input distribution **g**(ξ) premultiplied by constants and a quadratic phase factor.

Propagation to the Rear Focal Plane of a Converging Lens

The use of this relationship is again easiest if we use the previously defined matrices representing propagation through free space and passage through a thin lens. Recall that the matrix representing propagation through the lens when the medium preceding the lens has refractive index 1 is

```
In[]:= Clear[f, m3]
        m3[f_, 1] = {{1, 0}, {-1 / f, 1}}; MatrixForm[m3[f, 1]]
```

$$\begin{pmatrix} 1 & 0 \\ -\frac{1}{f} & 1 \end{pmatrix}$$

Consider an optical system consisting a region of free-space propagation (refractive index 1) of distance z and a thin lens with focal length f, followed by free-space propagation (refractive index 1) over distance f. Thus, we are examining the field in the rear focal plane of the thin lens. The ray-transfer matrix $\mathcal{M}_{tot}$ of the overall system can be found as the product of three matrices:

```
In[]:= mtot = m1[f, 1].m3[f, 1].m1[z, 1]
```

$$\left\{\{0, f\}, \left\{-\frac{1}{f}, 1 - \frac{z}{f}\right\}\right\}$$

We see that

```
In[•]:= 𝒜 = 0; ℬ = f; 𝒞 = -1 / f; 𝒟 = 1 - z / f;
```

Let $\mathbf{U}_1(\xi)$ represent the complex field in the input plane. The field $\mathbf{U}_2(y)$ represents the output field and can be found as

```
In[•]:= U2[y_] = FullSimplify[propagate[U₁[ξ]]]
```

$$\frac{e^{\frac{i \pi y^2 \left(1-\frac{z}{f}\right)}{\sqrt{z \lambda}}} \int_{-\infty}^{\infty} e^{-\frac{2 i \pi y \xi}{f \lambda}} U_1[\xi] \, d\xi}{\sqrt{i f \lambda}}$$

We see that the field distribution in the focal plane is a Fourier transform of the input function $\mathbf{U}_1[\xi]$ multiplied by a quadratic phase factor and scaled by constants. It is also obvious from the argument of the premultiplying exponential that if the distance z in front of then lens is equal to the focal length f of the lens, the quadratic phase factor preceding the Fourier transform vanishes, leaving only a scaled Fourier transform of the input field.

Propagation to an Image Plane

We now consider a geometry in which an object plane is distance z_1 in front of a positive lens of focal length f, and the field is observed in what we shall see, under the right conditions, is an image plane at distance z_2 behind the lens. The general form of the ray-transfer matrix for this geometry is

```
In[•]:= Clear[f]
        mtot = m1[z2, 1].m3[f, 1].m1[z1, 1];
        FullSimplify[MatrixForm[mtot]]
```

$$\begin{pmatrix} 1-\frac{z2}{f} & z1+z2-\frac{z1\,z2}{f} \\ -\frac{1}{f} & 1-\frac{z1}{f} \end{pmatrix}$$

where z_1 is the object distance from the lens and z_2 is the distance from the lens to where we look for an image to occur. We now apply the famous "lens law" relating z_1, z_2, and f that must be satisfied for an image to appear in the plane at distance z_2 beyond the lens:

$$\frac{1}{z_o} + \frac{1}{z_i} = \frac{1}{f}, \quad \text{or}$$

```
In[•]:= f = z1 * z2 / (z1 + z2);
```

Inserting this expression for f in the matrix representing the total system, we obtain

```
In[•]:= FullSimplify[mtot]; MatrixForm[FullSimplify[mtot]]
```

$$\begin{pmatrix} -\frac{z2}{z1} & 0 \\ -\frac{z1+z2}{z1\,z2} & -\frac{z1}{z2} \end{pmatrix}$$

```
In[•]:= 𝒜 = -z2 / z1; ℬ = 0; 𝒞 = - (z1 + z2) / z1 * z2; 𝒟 = -z1 / z2;
```

The matrix element $\mathcal{A}$ specifies the relationship between the coordinate ξ in the object plane and the coordinate y in the image plane. The negative sign indicates that the image is inverted, and the ratio z_2/z_1 is the magnification (or de-magnification) $\mathbb{M}$ of the system. The matrix element $\mathcal{D}$ specifies the relationship between the reduced angle $\hat{\theta}_1$ in the object space and the reduced angle $\hat{\theta}_2$ in the image space. Note that the reduced angle is changed in an inverse manner to that of the image size. That is, if the image is larger than the object by magnification $\mathbb{M}$, the reduced angle at the image is smaller than the reduced angle in the object space, the reduction factor being $1/\mathbb{M}$.

The fact that the matrix element $\mathcal{B}$ is zero has an important effect on the result of the **propagate[]** operator defined above.

```
In[ ]:= Quiet[Assuming[z1 ∈ Reals && z2 ∈ Reals && λ ∈ Reals && λ > 0 && y ∈ Reals,
          propagate[U1[ξ]]]]

        ComplexInfinity
```

We find an infinite result. Further investigation is needed. To start, we restate the original expression for the image field $\mathbf{U}(y)$, temporarily retaining $\mathcal{B}$ as non-zero:

$$\mathbf{U}\,(y;\,z_i) \;=\; \frac{e^{i\frac{2\pi L_0}{\lambda}}}{\sqrt{i\mathcal{B}\lambda}}\int_{-\infty}^{\infty}\mathbf{U}\,(\xi;\,0)\exp\left[\frac{i\,\pi}{\mathcal{B}\lambda}\left(\mathcal{A}\,\xi^2-2\,y\,\xi+\mathcal{D}\,y^2\right)\right]d\xi.$$

Now we replace $\mathcal{A}$ by $-\mathbb{M}$, $\mathcal{D}$ by $-1/\mathbb{M}$, and ignore the exponential in L_0 since it represents a constant phase, yielding

$$\mathbf{U}(y;\,z_i) \;=\; \frac{1}{\sqrt{i\mathcal{B}\lambda}}\int_{-\infty}^{\infty}\mathbf{U}(\xi;\,0)\exp\left[-\frac{i\,\pi}{\mathcal{B}\lambda}\left(\mathbb{M}\xi^2+2\,y\,\xi+\frac{y^2}{\mathbb{M}}\right)\right]d\xi.$$

This equation is a linear relation between the object complex field and the image complex field. We find the impulse response of this linear system, again temporarily retaining $\lambda\mathcal{B}$ as non-zero.

```
In[ ]:=
        Assuming[λℬ ∈ Reals && λℬ ≠ 0, 1 / Sqrt[I * λℬ] * Integrate[
          DiracDelta[ξ] * Exp[(-I * π / (λℬ)) * (M * ξ^2 + 2 * y * ξ + y^2 / M)],
          {ξ, -Infinity, Infinity}]]
```

$$\frac{e^{-\frac{i\,\pi\,y^2}{M\,\lambda\mathcal{B}}}}{\sqrt{i\,\lambda\mathcal{B}}}$$

Note that we are treating $\lambda\mathcal{B}$ as a single symbol since the two characters always appear together. As a matter of curiosity, we find the Fourier transform of this impulse response (i.e., the transfer function of this linear transformation).

```
In[ ]:= Assuming[λℬ ∈ Reals && λℬ ≠ 0 && M ∈ Reals && M > 0,
          ft1[Exp[((-I π y^2) / (M λℬ))] / Sqrt[I λℬ]]]
```

$$\frac{e^{i\,\pi\,M\,\lambda\mathcal{B}\,v^2}\sqrt{-i\,M\,\lambda\mathcal{B}}}{\sqrt{i\,\lambda\mathcal{B}}}$$

Canceling common terms in the numerator and denominator, we have

$$\mathcal{H}(v) = i\sqrt{\mathbb{M}}\, e^{i\pi\lambda\mathcal{B}(\sqrt{M}\,v)^2}.$$

From this result, we know that the value of the area under the impulse response is $\sqrt{\mathbb{M}}$; the value at the origin of the Fourier transform is equal to the area under the function that was transformed (it's easy to prove). We can drop the multiplicative factor i in the result, since it corresponds to a constant phase shift. Taking the limit as $\lambda\mathcal{B}$ goes to 0, it follows that

$$\lim_{\lambda\mathcal{B}\to 0} \mathcal{H}(v) = \sqrt{\mathbb{M}}.$$

This result is recognized as the Fourier transform of a delta function with area $\sqrt{\mathbb{M}}$. Thus, as $\lambda\mathcal{B} \to 0$, the impulse response $\mathbf{h}(y)$ approaches

$$\mathbf{h}(y) \to \frac{1}{\sqrt{\mathbb{M}}}\,\delta\left(\frac{-y}{\sqrt{\mathbb{M}}}\right),$$

where $-y$ is used to account for image inversion. Given an object complex field $\mathbf{U}_o(x)$, the resulting image field $\mathbf{U}_i(y)$ is found with a convolution of the impulse response with the object field. Evaluating the convolution by integration, we have

```
In[•]:= Assuming[M ∈ Reals && M > 0 && y ∈ Reals,
  Integrate[1 / (Sqrt[M]) * DiracDelta[-y/√M - x] * Uo[x],
   {x, -Infinity, Infinity}]]
```

$$\frac{\mathrm{U_o}\left[-\frac{y}{\sqrt{\mathbb{M}}}\right]}{\sqrt{\mathbb{M}}}$$

We conclude that when $\lambda\mathcal{B} \to 0$, the image complex field $\mathbf{U}_i(y)$ approaches $\frac{1}{\sqrt{\mathbb{M}}}\mathbf{U}_o\left(\frac{-y}{\sqrt{\mathbb{M}}}\right)$, a magnified and inverted version of the object complex field. It follows that the image intensity is given by

$$I_i(y) = \frac{1}{\mathbb{M}} I_o\left(-\frac{y}{\sqrt{\mathbb{M}}}\right),$$

where $I_o = |\mathbf{U}_o|^2$.

The reader may wonder why the image complex field has no multiplicative exponential quadratic-phase factors as are found in other analyses (see Ref. [4], Section 6.3). The reason is that we have made the unrealistic but simplifying assumption that the lens is infinite in size. The system therefore maps a point source into a point image, with no diffraction effects. In practice, the lens has a finite exit pupil, which modifies the analysis substantially. A more realistic analysis, including a finite lens pupil, requires propagation from the object plane through the lens to the lens exit pupil, imposition of the finite pupil on the resulting field, and then propagation to the image plane. The result is considerably more complicated than the result shown here. See Ref. [6] for details.

6. Discrete Diffraction Calculations

There are many cases in which a purely symbolic calculation of a diffraction pattern of an aperture is not possible. The integrals involved may be too complicated for *Mathematica®* to find a solution, or the description of the aperture itself may be difficult to state in a purely analytical form. In such cases, one must resort to discrete approximations to both the diffracting aperture and the diffraction equations themselves. In such calculations, continuous functions are replaced by sampled function arrays, and integrals are replaced by finite summations. Discrete diffraction calculations are the subject of this chapter. A much more detailed discussion of this subject can be found in Ref. [4], Chapter 5.

6.1 The Discrete Fourier Transform (DFT)

As we have seen, the Fourier transform plays a central role in continuous diffraction calculations. For discrete versions of such calculations, it is the discrete Fourier transform that plays the equivalent role. We therefore begin with definitions of the discrete Fourier transform and its inverse, both in one dimension. For more detailed information regarding the properties of the discrete Fourier transform, see Ref. [1], Chapter 10.

The discrete Fourier transform will be represented by an operator $\mathcal{DFT}\{\}$. Its definition is given by

$$\mathbf{V}_n = \mathcal{DFT}\{\mathbf{U}_k\} \triangleq \sum_{k=0}^{N-1} \mathbf{U}_k \exp\left(-i\,2\pi\frac{k\,n}{N}\right), \qquad n = 0, \cdots, N-1.$$

An important property of the DFT is that it is periodic with period N, i.e., $\mathbf{V}_{q\,N+p} = \mathbf{V}_p$ for any integers p, q. In addition, the spectral sequence $\mathbf{V}_n$ is ordered such that the positive frequency components are represented by indices $n = 0, \cdots, N/2 - 1$, and the negative frequencies by indices $N/2, \cdots, N-1$, assuming that N is an even number. Usually N is chosen to be a power of 2 in order to implement the simplest form of a fast algorithm called the fast Fourier transform (again, see Ref. [1], Chapter 10).

The inverse of the DFT is represented by the operator $\mathcal{DFT}^{-1}\{\}$ and is given by

$$\mathbf{U}_k = \mathcal{DFT}^{-1}\{\mathbf{V}_n\} \triangleq \frac{1}{N}\sum_{n=0}^{N-1} \mathbf{V}_n \exp\left(+i\,2\pi\frac{k\,n}{N}\right), \qquad k = 0, \cdots, N-1.$$

This definition assures that $\mathcal{DFT}^{-1}\{\mathcal{DFT}\{\mathbf{U}_k\}\} = \mathbf{U}_k$.

In *Mathematica*, the discrete Fourier transform is carried out by the command `Fourier[`*arg*`,FourierParameters->{a,b}]`, where *arg* is an array or list of numbers (real or complex) of any dimension, and the parameters `a` and `b` are chosen to achieve the form of the DFT presented above. The choice of parameters for our definition is `{1,-1}`. The command to perform the inverse DFT in *Mathematica* is `InverseFourier[`*arg*`,FourierParameters->{a,b}]`, where *arg* is now the spectral array (of any dimension) we wish to subject to an inverse DFT. Again the parameter choice to match our definition is `{1,-1}`.

One further subtlety you should be aware of is that *Mathematica* always numbers the elements in a list or array 1 through N rather than 0 though $N-1$. The definitions of these two operations in *Mathematica* take account of this numbering scheme, but if you wish to select a particular element of the spectral sequence, you should remember this fact.

Finally, the reader should recognize the following well-known properties of the DFT. First, it is a basic property of the DFT that the product of the sampling interval $\Delta\xi$ in the space domain and the sampling interval $\Delta\nu$ in the frequency domain is equal to $1/N$; that is, $\Delta\xi\,\Delta\nu = 1/N$. Second, for fixed sampling interval $\Delta\xi$, the larger one chooses the padded aperture sequence length N, the smaller the sampling increment $\Delta\nu$ in the frequency domain, by virtue of the fact that $\Delta\nu = 1/(\Delta\xi\, N)$. Thus, if the spectrum has a lot of fine detail, one needs a large N to capture that detail. In addition, as M, the number of samples falling within the aperture itself, is decreased, two phenomena happen. First, the representation of the aperture function becomes coarser and less accurate. Second, the Fourier transform becomes wider.

In Chapter 1, we defined two operators that perform the discrete Fourier transform and its inverse. These operators are `dft[]` and `idft[]`. They are defined in this chapter in a hidden cell near the start of this chapter, and we shall use these operators in the examples that follow.

6.2 Discrete Diffraction Equations

Let the symbols Δx and $\Delta\xi$ represent the sample spacings in the observation plane and the aperture plane, respectively. How to choose these parameters will be discussed in Section 6.3. For the moment, our goal is to express the Fresnel transform form and the Fresnel transfer function form of the discrete diffraction formulas.

Fresnel Transform Approach

To start, we repeat the continuous 1D version of the Fresnel transform form of the Fresnel diffraction equation,

$$\mathbf{U}(x;\, z) = \frac{e^{i\frac{\pi}{\lambda z}x^2}}{\sqrt{i\,\lambda\, z}} \int_{-\infty}^{\infty} \left[\mathbf{U}(\xi;\, 0)\, e^{i\frac{\pi}{\lambda z}\xi^2}\right] e^{-i\frac{2\pi}{\lambda z}\xi\, x}\, d\xi.$$

Again, $\mathbf{U}(x;\, z)$ represents the diffracted field at distance z, while $\mathbf{U}(x;\, 0)$ represents the field in the plane of the diffracting aperture.

We focus initially on finding an expression for the sample spacing $\Delta\xi$ in terms of the finite width of the aperture and the number of samples within that width. Let the width of the finite aperture be represented by ℓ, and the number of samples lying within the aperture by M. It follows directly then that $\Delta\xi = \ell/M$.

Finding Δx is a bit more tricky. This solution depends on a relationship mentioned above that holds for any DFT, namely,

$$\Delta\xi\,\Delta\nu = 1/N,$$

where $\Delta\nu$ is the sample spacing in the frequency domain and N is the total number of samples in the calculation (note, $N \geq M$). In the present case, the frequency variable is given by $\nu = x/(\lambda z)$. Therefore, the following relation must hold:

$$\Delta\xi\,\Delta x = \lambda z/N.$$

Thus,

$$\Delta x = \frac{\lambda z}{\Delta\xi\, N} = \frac{M}{N}\frac{\lambda z}{\ell} = \frac{1}{Q}\frac{\lambda z}{\ell},$$

where $Q = N/M$ is termed the sampling ratio. Note that for fixed Q, ℓ, and λ, the spacing Δx between samples in the observation plane increases with z.

Now we are in a position to write a discrete form of the 1D Fresnel diffraction formula. Making the substitutions

$\xi = m\Delta\xi = \frac{m}{M}\ell$, where m runs from 0 to $M-1$, and $x = n\,\Delta x = \frac{n\,M}{N}\frac{\lambda z}{\ell}$, where n runs from 0 to $N-1$, we obtain for the nth sample value $\mathbf{U}_n(z)$ observed at distance z,

$$\mathbf{U}_n(z) = \exp\left[i\frac{\pi}{4\,Q^2\,N_F}(n-N/2)^2\right]\times\sum_{k=0}^{N-1}\mathbf{U}_k(0)\exp\left[i\,\pi\frac{4\,N_F}{M^2}(k-N/2)^2\right]\exp\left(-i\frac{2\,\pi}{N}\,n\,k\right),$$
$$n = 0,\cdots,N-1,$$

where we have substituted the Fresnel number N_F for $\frac{(\ell/2)^2}{\lambda z}$. The result can equivalently be expressed as a 1D DFT,

$$\mathbf{U}_n(z) = \exp\left[i\frac{\pi}{4\,Q^2\,N_F}(n-N/2)^2\right]\mathcal{DFT}\left\{\mathbf{U}_k(0)\exp\left[i\,\pi\frac{4\,N_F}{M^2}(k-N/2)^2\right]\right\}.$$

Generalization to a 2D form of the equation is straightforward for an aperture that is enclosed within a circle of diameter ℓ:

$$\mathbf{U}_{n,m}(z) = \exp\left[i\frac{\pi}{4\,Q^2\,N_F}\left[(n-N/2-1)^2+(m-N/2-1)^2\right]\right]\times$$
$$\sum_{k=1}^{N}\sum_{p=1}^{N}\mathbf{U}_{k,p}(0)\times\exp\left[i\,\pi\frac{4\,N_F}{M^2}\left[(k-N/2-1)^2+(p-N/2-1)^2\right]\right]\times$$
$$\exp\left[-i\frac{2\,\pi}{N}((n-1)(k-1)+(m-1)(p-1))\right],$$
$$n = 1,\cdots,N,\quad m = 1,\cdots,N.$$

However, again keep in mind that *Mathematica* numbers lists (arrays) with indices 1 through N rather than 0 through $N-1$. Therefore, the proper 2D Fresnel diffraction equation in *Mathematica* notation and with the definition of the DFT adopted here is

$$\mathbf{U}_{n,m}(z) = \exp\left[i\frac{\pi}{4\,Q^2\,N_F}\left[(n-N/2-1)^2+(m-N/2-1)^2\right]\right]\times$$
$$\sum_{k=1}^{N}\sum_{p=1}^{N}\mathbf{U}_{k,p}(0)\exp\left[i\,\pi\frac{4\,N_F}{M^2}\left[(k-N/2-1)^2+(p-N/2-1)^2\right]\right]\times$$
$$\exp\left[-i\frac{2\,\pi}{N}((n-1)(k-1)+(m-1)(p-1))\right],$$
$$n = 1,\cdots,N,\quad m = 1,\cdots,N.$$

The double summation is a 2D DFT of the sampled aperture function times a quadratic phase function:

$$\mathbf{U}_{n,m}(z) = \exp\left[i\frac{\pi}{4\,Q^2\,N_F}\left[(n-N/2-1)^2+(m-N/2-1)^2\right]\right]\times$$
$$\mathcal{DFT}\left\{\mathbf{U}_{k,p}(0)\exp\left[i\,\pi\frac{4\,N_F}{M^2}\left[(k-N/2-1)^2+(p-N/2-1)^2\right]\right]\right\}.$$

There are three important insights to be drawn from the above discussion:

1) From the expression for Δx, we see that, for fixed Q, ℓ, and λ, the spacing of samples in the observation plane increases as z increases. But in addition, we know that as z increases, eventually the width of the physical diffraction pattern also increases (Fraunhofer diffraction). These two simultaneous increases result in the width of the discrete calculation of the diffraction pattern staying more or less constant as z increases. This behavior is different than that found when the Fresnel transfer function approach is used for the calculation, as discussed in the next section.

2) As can be seen from the factor $\frac{4N_F}{M^2}$ in the argument of the quadratic-phase exponential under the summation signs in both the 1D and 2D cases, for a fixed M, that exponent has increasing oscillations as N_F increases. This fact implies that the bandwidth of the exponential increases with increasing N_F or, equivalently, the bandwidth of the exponential increases as we move deeper and deeper into the Fresnel region of diffraction.

3) From the same factor $\frac{4N_F}{M^2}$, we note that for a fixed N_F, the bandwidth of the quadratic-phase exponential within the DFT decreases as we increase the number of samples M within the diffracting aperture.

We now turn our attention to an alternative method for calculating diffraction patterns.

Fresnel Transfer Function Approach

In Chapter 3 we showed that the 1D Fresnel transfer function approach in the continuous case can be represented by the equation

$$\mathbf{U}(x; z) = \mathcal{F}^{-1}\{\mathcal{F}[\mathbf{U}(x; 0)] \times e^{ikz} \exp[i\pi\lambda z \nu_x^2)]\},$$

in which the script $\mathcal{F}$ represents a Fourier transform operator and $\mathcal{F}^{-1}$ represents its inverse. Thus, we have multiplied the Fourier transform of the aperture distribution by the Fresnel transfer function of free-space propagation, and inverse-transformed the product. In what follows, we neglect the term e^{ikz} since it presents a constant phase shift that does not affect the intensity of the result.

The sample spacings in this approach differ from those for the Fresnel transform approach. Let the total number of samples in the calculation be N, and let the total width of the aperture with padding be L. The spacing of the samples in x is $\Delta x = L/N$. Now the discrete Fourier transform requires that $\Delta x\,\Delta\nu_x = 1/N$. It follows that the spacing of samples in the frequency domain must be $\Delta\nu_x = 1/(N\Delta x) = 1/L$. The aperture itself has diameter ℓ and contains M samples. It follows that $M = \frac{\ell}{L}\,N$.

From this point on, we proceed with an approximate analysis that relates the normalized slope $\frac{1}{2\pi}d/d\nu_x$ of the transfer function exponential to position $\tilde{x}$ in the space domain, namely,

$$\tilde{x} = \frac{1}{2\pi}\frac{d}{d\nu_x}\left(\pi\lambda z\nu_x^2\right) = \lambda z\,\nu_x.$$

While the transfer function has significant value for all values of ν_x, the width of the total spectrum is governed by the effective bandwidth B_x of the aperture function, which after sampling is given by $B_x = \frac{1}{\Delta x} = \frac{M}{\ell}$. Given the above expression for $\tilde{x}$, this implies that the length of the fully padded sequence should be

$$L = \lambda z\,B_x = \lambda z\frac{M}{\ell}.$$

If we solve this equation for M, we obtain

$$M = \frac{L\ell}{\lambda z} = 4QN_F.$$

It then follows that the total number of samples, with padding, should be

$$N = Q\,M = 4\,Q^2\,N_F.$$

The values of N, M, and Q must be found by trial and error, as described in the section that follows. The transfer function of free-space propagation can now be written in discrete form as follows:

$$\mathcal{H}\,(k\,\Delta\nu_x) = \exp\left[-i\,\pi\,\lambda z\left(\frac{k-N/2}{L}\right)^2\right] = \exp\left[-i\,\pi\left(\frac{\ell}{L}\right)^2\frac{(k-N/2)^2}{4\,N_F}\right] =$$

$$\exp\left[-i\,\frac{\pi}{4\,Q^2\,N_F}\,(k-N/2)^2\right].$$

It follows that the discrete diffraction formula for the Fresnel transfer function approach is given by

$$\mathbf{U}_n(z) = \mathcal{DFT}^{-1}\left\{\mathcal{DFT}\{\mathbf{U}_k(0)\}\exp\left[-i\,\frac{\pi}{4\,Q^2\,N_F}\,(k-N/2)^2\right]\right\},$$

where $\mathcal{DFT}\{\}$ represents a discrete Fourier transform while $\mathcal{DFT}^{-1}\{\}$ is an inverse discrete Fourier transform. A 2D version of the result can be written as

$$\mathbf{U}_{n,m}(z) = \mathcal{DFT}^{-1}\left\{\mathcal{DFT}[\mathbf{U}_{k,p}(0)]\times\exp\left[-i\,\frac{\pi}{4\,Q^2\,N_F}\,\left((k-N/2)^2+(p-N/2)^2\right]\right\} =$$

$$\mathcal{DFT}^{-1}\left\{\mathcal{DFT}[\mathbf{U}_{k,p}(0)]\times\exp\left[-i\,\frac{\pi\,M^2}{4\,N^2\,N_F}\,\left((k-N/2)^2+(p-N/2)^2\right]\right\}.$$

Some general observations about this result follow:

1) The larger the Fresnel number N_F, the slower the phase oscillations of the quadratic-phase exponential transfer function. This behavior is the opposite of the behavior observed for the Fresnel transform approach given above.

2) The smaller the M value and the larger the N value, the slower the phase oscillations of the transfer function. Thus, a large amount of zero padding of the aperture function is beneficial. However, while it might appear that a small value of M is desirable, M must still be large enough to properly sample the variations of the aperture function.

It remains to specify the values of N, M, and Q required for accuracy of these diffraction formulas.

6.3 Sampling Requirements

Creating a discrete version of a continuous or symbolic aperture function requires a choice of the sample spacing in two dimensions. The sample spacing required for accuracy in the diffraction pattern is determined by the sampling theorem, which is discussed in detail in Chapter 8 of Ref. [1]. To summarize that theorem, if the total widths of the Fourier transform of the integrand of the diffraction integral in the ν_X and ν_Y frequency directions are represented by B_X and B_Y, then the sample spacings in the integrand should be no larger than $1/B_X$ and $1/B_Y$. If the integrand is perfectly bandlimited to this region of frequency space, then the original function can be perfectly recovered by convolving the samples with an interpolation function of the form $\text{sinc}(B_X\,x)\times\text{sinc}(B_Y\,y)$.

In practice, aperture functions are always finite in extent. It can be shown that this fact implies that the spectrum of an aperture function can never be perfectly bandlimited, although it can be approximately so. Departures from band limitation imply that there will always be some degree of aliasing (see Chapter 8 of Ref. [1] for a more detailed discussion of sampling and aliasing). Thus, in discrete diffraction calculations, the sampling interval must be chosen small enough to assure that the aliasing errors at the edge of the spectrum are small enough to be tolerable.

If one is using a discretized version of the Fresnel diffraction integral in its Fresnel transform form, one must be sure to adequately sample the product of the aperture function and the quadratic-phase exponential. Which of these two components dominates the sampling requirements depends on the Fresnel number N_F, defined previously as

$$N_F = \frac{(\ell/2)^2}{\lambda z}.$$

When $\lambda z \gg (\ell/2)^2$, $N_F \ll 1$ and the bandwidth is dominated by the aperture function. When $\lambda z \ll (\ell/2)^2$, then $N_F \gg 1$ and the bandwidth is dominated by the quadratic-phase exponential.

Note that there are some aperture functions, such as a truncated Gaussian, for which the amplitude of the aperture field may become small before reaching the maximum dimension ℓ of the aperture. In such cases, one can define ℓ as an effective width, perhaps using the equivalent width [1] of the aperture function, or possibly some other measure of width.

There is no known analytical method to determine the maximum allowable sampling interval for an arbitrary aperture function and a given Fresnel number, in particular one that simultaneously guarantees a desired level of aliasing at the edge of the resulting discrete diffraction pattern. However, it is possible to choose a starting point in an iterative process that adjusts the sampling rate depending on the aliasing that results for each choice of sampling interval. A starting point in such an iterative process might be provided by the results obtained for a uniform rectangular aperture, with a multitude of trial-and-error calculations aimed at yielding a pre-ordained level of aliasing at the edge of the diffraction pattern intensity.

The question remains as to what is a reasonable level of aliasing error to allow. The answer depends on the requirements of the user and the method chosen for display of the results. Most commonly, it is the intensity distribution in the diffraction pattern that is of interest and displayed for the user. Some reasonable goals might be intensity errors on the order of 10^{-2} relative to the central intensity peak if the diffraction pattern is displayed on a linear scale, and perhaps 10^{-4} relative to the central intensity peak if the pattern is displayed on a log scale.

The level of aliasing is determined by the amount of zero-padding given to the aperture field distribution. In what follows, we again designate the number of samples required within the aperture by the symbol M, and the total number of samples after zero-padding by the symbol N. The values of both N and M depend on the method used for calculating the diffraction pattern. The Fresnel transform method requires a single Fourier transform, which is carried out using the fast Fourier transform (FFT) algorithm, while the transfer function method requires a Fourier transform of the aperture function, multiplication by the transfer function of free-space propagation, and an inverse Fourier transform.

To provide a starting point in an iterative procedure for finding the required values of N and M for a given aperture and a given level of aliasing, the values required for a rectangular aperture can be used. These values are presented in Ref. [4], Chapter 5, derived from exhaustive simulations for various values of Fresnel number N_F. The reader is referred to this reference to see the details. We choose the parameters M and N in the examples based on the curves in that reference.

6.4 Some Discrete Diffraction Examples With *Mathematica*

Circular Aperture Functions

In Section 3.7, several diffraction patterns for a circular aperture and various Fresnel numbers were presented. Here we use the discrete Fresnel transform approach to calculate one of these patterns, allowing visual comparison of the results from a continuous calculation and a discrete calculation.

Our first task is to create a discrete version of a circular aperture. We choose the size of the $M \times M$ array by choosing the value of M. This can be done with *Mathematica* commands as follows:

```
In[ ]:= Clear[m];
        m = 64;
        aperture = Table[circ[√(i - 1 - m / 2) ^2 + (j - 1 - m / 2) ^2 / (m / 4)],
           {i, 1, m}, {j, 1, m}];
        ListDensityPlot[aperture]
```

Out[ ]=

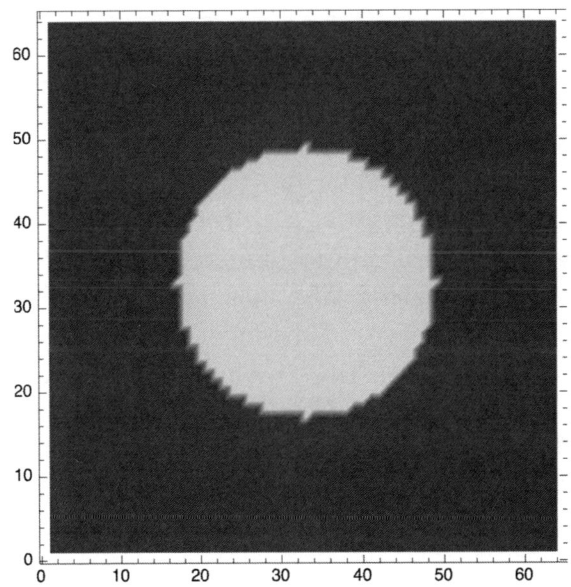

Note the use of the lower-case letter **m** rather than an upper-case *M*. In *Mathematica*, it is always good practice to use lower-case letters for variables in order to avoid possible confusion with an upper-case variable that is defined as part of *Mathematica* itself. The argument **m/4** defines the radius of the circle in pixels, and the diameter is therefore **m/2**, or 32 in this case.

Note also the ragged edges of the circle, a consequence of attempting to approximate a circularly symmetric array with a rectilinear set of samples. A much better representation occurs if we choose **m** to be larger (with the price of longer computation time):

```
In[ ]:= Clear[m];
        m = 256;
        aperture = Table[circ[√(i - m / 2 - 1) ^2 + (j - m / 2 - 1) ^2 / (m / 4)],
           {i, 1, m}, {j, 1, m}];
        ListDensityPlot[aperture]
```

Out[ ]=

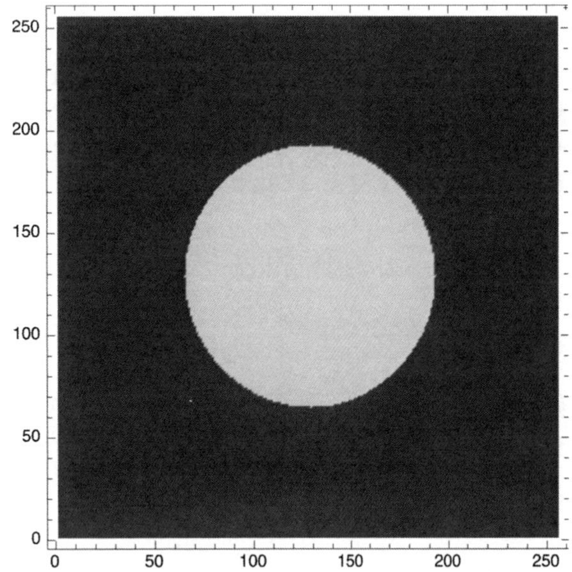

The ragged edges of the circle are much less prominent in this case.

Diffraction Calculation Using the Fresnel Transform

Next we must choose the values of the parameters N_F, M, and N and the degree of aliasing that is allowable. Referring again to the plots in Ref. [4], Chapter 5, we choose $N_F = 10$ because of the interesting structure of the diffraction pattern in this case, and we allow aliasing of the diffraction pattern at the level 10^{-2} of the central maximum value in the diffraction intensity pattern. With these assumptions, reasonable choices of M and N are 128 and 512, respectively (there are some signs of aliasing if we use the numbers 64 and 256, as might be guessed from the curves presented in Ref. [4]). Thus, we must pad the circular aperture with zeros such that the array representing the aperture is 512 × 512. Remember, however, that the plots in the figure are for a 1D rectangular aperture; for a circular aperture, which is more difficult to accurately represent in discrete form, larger sample numbers are likely needed. The *Mathematica* code below calculates an intensity distribution of the diffraction pattern by means of a discrete computation. First we define the aperture function and the Fresnel number:

```
In[ ]:= m = 128; n = 512; nf = 10;
aperture = Parallelize[
   Table[circ[Sqrt[(i - 1 - n / 2) ^2 + (j - 1 - n / 2) ^2] / (m / 2)],
    {i, 1, n}, {j, 1, n}]];
```

Next we define the quadratic-phase kernel. Note that we define it only within an $M \times M$ array, since such an array covers the circular aperture with diameter M.

```
In[ ]:= kernel = Parallelize[Table[
      Exp[I * Pi * ((4 * nf) / m^2) * ((i - 1 - m / 2) ^2 + (j - 1 - m / 2) ^2)],
      {i, 1, m}, {j, 1, m}]];
```

It is good practice to check if the kernel is actually sampled finely enough to avoid aliasing. To do so, we make density plots of the real and imaginary parts of the kernel and see if signs of aliasing are present. Note that only signs of aliasing within the dimensions of the circular aperture are important, for the product of the aperture function and the exponential will be zero outside the aperture. If the aperture is not perfectly space limited, then make sure there is no aliasing within some estimate of the effective width of the aperture function. Aliasing of this function generally shows up as circular moiré fringes near the edges of the pattern.

In[ ]:=
```
p1 = ListDensityPlot[Re[kernel],
    PlotLabel → "Real Part", ImageSize → 200];
p2 = ListDensityPlot[Im[kernel],
    PlotLabel → "Imaginary Part", ImageSize → 200];
GraphicsRow[{p1, p2}]
```

Out[ ]=

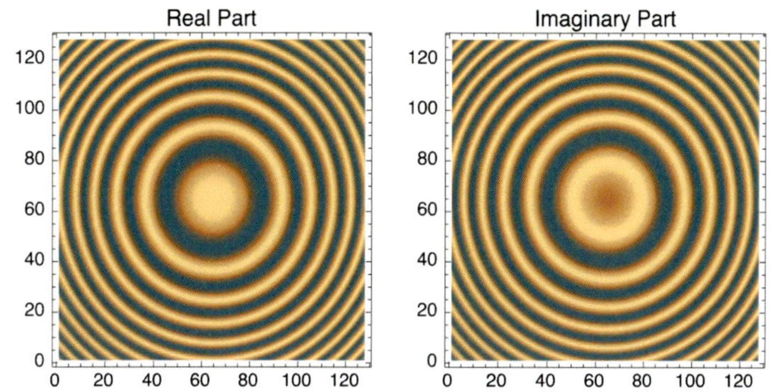

Note that for $N_F \gg 1$, it is the quadratic-phase function that dominates the bandwidth, so the most important place to look for aliasing is in the representation of that function. There are no moiré patterns evident, so we conclude that the sampling is fine enough.

Next we zero-pad the array to size $N \times N$ and move the exponential array to be centered at indices ($N/2$, $N/2$). To do so, we use the **PadRight[]** command followed by the **RotateRight[]** command, and then display the real part of the expanded **kernel1** array:

In[ ]:=
```
kernel1 =
   RotateRight[PadRight[kernel, {n, n}], {n / 2 - m / 2, n / 2 - m / 2}];
ListDensityPlot[Re[kernel1],
 PlotLabel → "Real Part of Padded Array"]
```

Out[ ]=

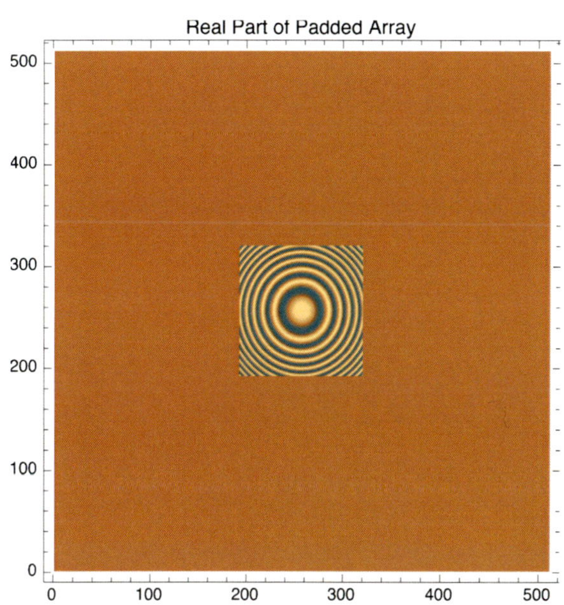

Now we find the amplitude distribution in the diffraction pattern by taking the DFT of the product of the aperture array and the quadratic-phase exponential array. Finally, we display a portion of a central slice through the normalized intensity distribution of the diffraction pattern.

```
In[•]:= amplitude = dft[aperture * kernel1];
        intensity = RotateRight[Abs[amplitude]^2, {n/2, n/2}];

In[•]:= ListPlot[Take[RotateRight[intensity[[n/2]] / Max[intensity], n],
          {n/2 - m, n/2 + m}], PlotRange → All, Joined → True,
         Filling → Bottom, AxesLabel → {"Pixel Number", I}]
```

Out[•]=

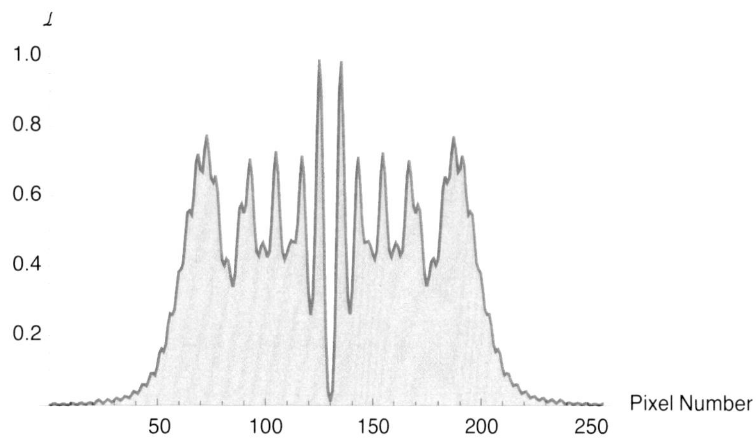

This distribution appears to be an accurate result when compared with the same case found by numerical integration of the Hankel transform shown in Section 3.7.

Diffraction Calculation Using the Fresnel Transfer Function

In this second example, we use the Fresnel transfer function approach to calculate the diffraction pattern of an aperture that is again the circular aperture of diameter M, using the same value $N_F = 10$ as used above. We experiment to find the best values of M and N, again referring to Ref. [4], and settling on $M = 128$ and $N = 512$.

Using *Mathematica*'s sequence numbering system, the Fresnel transfer function sequence is given by

$$H_{k,p} = \exp\left[-i\,\frac{\pi M^2}{4\,N^2\,N_F}\left((k - N/2 - 1)^2 + (p - N/2 - 1)^2\right)\right], \quad k = 1, \cdots, N, \;\; p = 1, \cdots, N.$$

A good test of the accuracy of the calculation is to check whether the intensity of the diffraction pattern falls to zero at the center of the pattern, as we have seen to be the case in the figure in the previous subsection. We try these numbers in the following. After the calculation, a slice through the diffraction pattern intensity is taken through its center:

```
In[•]:= m = 128; n = 512; nf = 10;
        aperture = Parallelize[
           Table[circ[Sqrt[(i - 1 - n/2)^2 + (j - 1 - n/2)^2] / (m/2)],
            {i, 1, n}, {j, 1, n}]];
        spectrum = RotateRight[dft[aperture], {(n/2), n/2}];
```

```
    h = Parallelize[Table[
    Exp[-I * π * (m^2 / (4 * n^2 * nf)) * ((k - n / 2 - 1) ^2 + (p - n / 2 - 1) ^2)]
    {k, 1, n}, {p, 1, n}]];
amplitude = idft[spectrum * h];
intensity = Abs[amplitude] ^2;
ListPlot[Take[intensity[[n / 2]] / Max[intensity], {145, 360}],
 Joined → True, Filling → Bottom, AxesLabel → {"Pixel Number", I}]
```

Out[ ]=

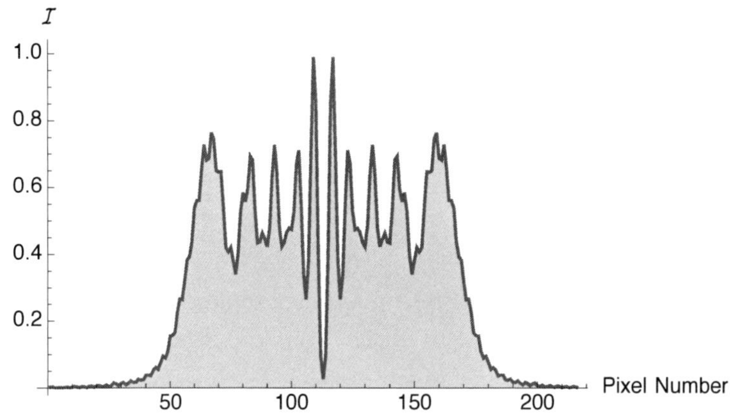

As can be seen from this result, the value at the center of the pattern does not quite fall all the way to zero, partly because the discretized circle has irregular edges and partly because the padded array is not large enough.

Before going further, a few words about the above program are in order. The first line sets the starting parameters. The second line defines the discrete circular aperture. The third line calculates the DFT of that aperture and rotates the results to be centered at indices ($N/2$, $N/2$). The fourth line calculates the Fresnel transfer function, also centered at ($N/2$, $N/2$). The fifth line calculates the spectrum of the circular aperture times the Fresnel transfer function and performs an inverse DFT on that product. The corresponding intensity is calculated in the sixth line. In the last line, a 1D slice though the center of the pattern is taken and displayed over a limited range of indices.

Experimentation shows that good results for this aperture and this method of calculation occur for $M = 256$ (yielding a more accurate representation of the circular aperture) and $N = 2048$ (yielding slower variations of the phase of the quadratic-phase exponential), as shown below (be patient; the calculation takes some time):

```
In[ ]:= m = 256; n = 2048; nf = 10;
aperture = Parallelize[
   Table[circ[Sqrt[(i - 1 - n / 2) ^2 + (j - 1 - n / 2) ^2] / (m / 2)],
    {i, 1, n}, {j, 1, n}]];
spectrum = RotateRight[dft[aperture], {(n / 2), n / 2}];
h = Parallelize[Table[
    Exp[-I * π * (m^2 / (4 * n^2 * nf)) * ((k - n / 2 - 1) ^2 + (p - n / 2 - 1) ^2)]
    {k, 1, n}, {p, 1, n}]];
amplitude = idft[spectrum * h];
intensity = Abs[amplitude] ^2;
ListPlot[Take[intensity[[n / 2]] / Max[intensity], {800, 1250}],
 Joined → True, Filling → Bottom, AxesLabel → {"Pixel Number", I}]
```

Out[ ]=

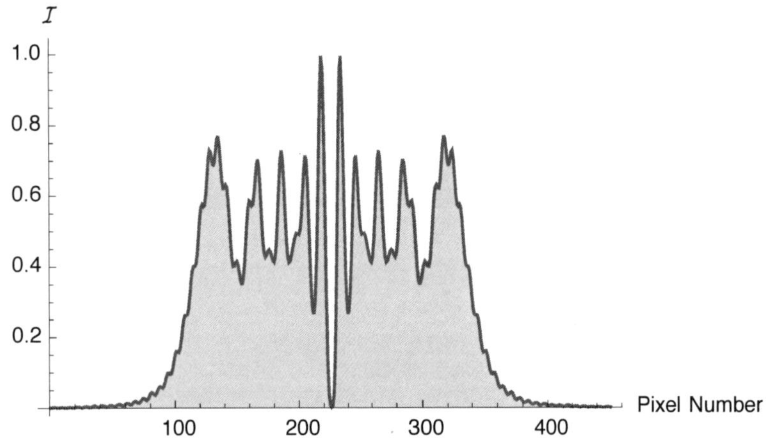

In this result, the value of the diffraction pattern intensity goes all the way to zero at the center of the pattern. Smaller values of M and N do not yield a zero at the center. The pixel number scales are different in the two figures because of finer sampling in the second calculation.

We have seen that both the Fresnel transform method and the Fresnel transfer function method can be used to calculate the diffraction pattern with good accuracy, although in this case the transfer function method requires larger arrays for accurate results.

6.5 A Diffraction Pattern That Does Not Change Shape as the Fresnel Number Changes

As a final example, which is particularly interesting, we calculate a diffraction pattern that is independent of Fresnel number. The aperture function chosen is a thin ring of fixed radius, two pixels thick. The code below calculates that aperture, and the plot illustrates its shape. In this first calculation, $M = 64$, $N = 1024$:

In[ ]:=
```
m = 64; n = 1024;
aperture = Parallelize[
   Table[circ[Sqrt[(i - 1 - n / 2) ^2 + (j - 1 - n / 2) ^2] / (m / 2)] -
     circ[Sqrt[(i - 1 - n / 2) ^2 + (j - 1 - n / 2) ^2] / (m / 2 - 2)],
    {i, 1, n}, {j, 1, n}]];
Image[aperture]
```

Out[ ]=

We now set the Fresnel number to 10 and calculate the normalized intensity distribution in the diffraction pattern using the Fresnel transform approach, displaying a central slice through the result with second-order interpolation and normalizing by the maximum value of intensity found at the center of the pattern:

In[•]:=

```
nf = 10;
kernel = Parallelize[
   Table[Exp[I * Pi * ((4 * nf) / m^2) * ((i - 1 - n / 2) ^2 + (j - 1 - n / 2) ^2)],
    {i, 1, n}, {j, 1, n}]];
amplitude = dft[aperture * kernel];
intensity = RotateRight[Abs[amplitude] ^2, {n / 2, n / 2}];
p1 = ListPlot[Take[RotateRight[
    intensity⟦n / 2⟧ / Max[intensity⟦n / 2⟧], n], {420, 606}],
  PlotRange → All, Joined → True, InterpolationOrder → 2,
  AxesLabel → {"Pixel Number", I}, PlotLabel → "NF = 10"]
```

Out[•]=

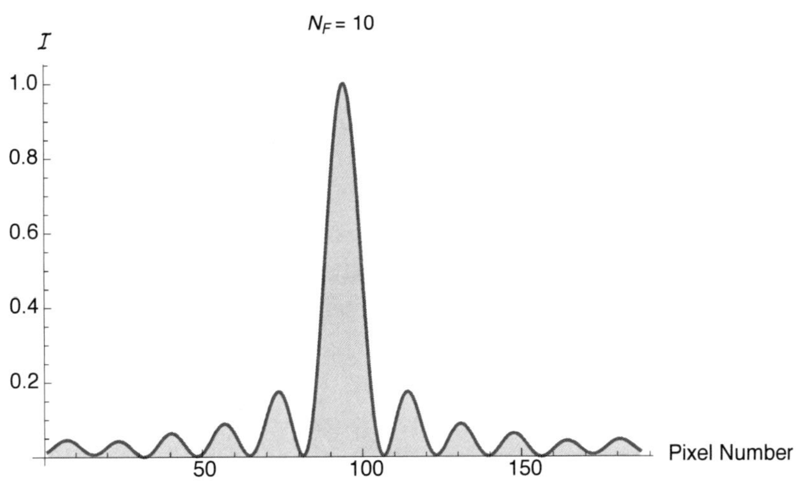

Keeping this result in mind, we repeat the calculation with the same values of M and N, but changing the Fresnel number by two orders of magnitude to $N_F = 0.1$:

In[•]:=

```
nf = 0.1;
kernel = Parallelize[
   Table[Exp[I * Pi * ((4 * nf) / m^2) * ((i - 1 - n / 2) ^2 + (j - 1 - n / 2) ^2)],
    {i, 1, n}, {j, 1, n}]];
amplitude = dft[aperture * kernel];
intensity = RotateRight[Abs[amplitude] ^2, {n / 2, n / 2}];
p2 = ListPlot[Take[RotateRight[
    intensity⟦n / 2⟧ / Max[intensity⟦n / 2⟧], n], {420, 606}],
  PlotRange → All, Joined → True, InterpolationOrder → 2,
  AxesLabel → {"Pixel Number", I}, PlotLabel → "NF = 0.1"]
```

Out[ ]=

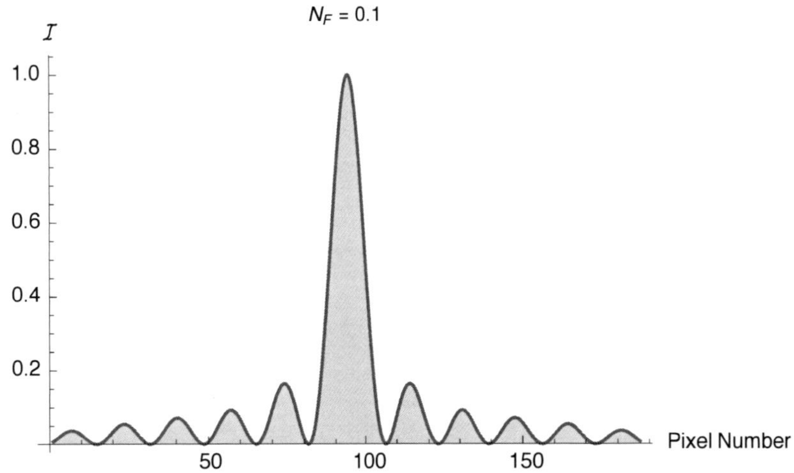

We see that the normalized intensity distribution is independent of the Fresnel number over a wide range of that number! For a ring aperture of fixed diameter, the change of Fresnel number is equivalent to changing the observation distance by two orders of magnitude. Keep in mind that the Fresnel transform approach increases the spacing of samples in the diffraction pattern as z is increased. So the diffraction pattern is actually expanding with increasing z, and as the pattern expands, the unnormalized height of the diffraction pattern decreases, but its shape is not changing.

The explanation for this behavior is as follows. Since the aperture function and the quadratic-phase exponential are multiplied together, the thin-ring aperture simply samples the quadratic-phase exponential, multiplying the aperture function by a constant phase factor, the phase of which depends on the radius of the ring. When we calculate the intensity of the diffraction pattern, that constant phase multiplier is immaterial. Thus, the diffraction pattern is always the squared magnitude of the DFT of the aperture function, and in normalized form does not change with changes in the Fresnel number. We should note, however, that for Fresnel numbers larger than about 10, the results depart from what has been shown. It would be necessary to increase N and M to obtain accurate results for larger Fresnel numbers, as implied by the curves in Section 5.4 of Ref. [4].

7. Imaging With Coherent Light

In this chapter, we consider simulation of fully coherent imaging systems. We build models for such systems and explore the properties of the images they form. In the chapter that follows this one, we explore the corresponding properties of imaging systems that use fully incoherent light.

7.1 Coherent vs. Incoherent Imaging

In this section, we give a brief explanation of the difference between coherent and incoherent imaging. For additional material on this subject, see Ref. [7], Section 7.4, and for a general discussion of coherence, See Ref. [8]. The difference between these two modes of imaging lies in the nature of the object illumination in each case.

For coherent imaging, the illumination at all points on the object may have various amplitudes and phases, but the sinusoidal fluctuations of the optical field have phases that are locked together. Thus, at one point on the object, the amplitude of the light may be represented by a certain complex number (a phasor), while at another point on the object, it may be represented by another complex number (another different phasor), but the crucial property of full coherence is that those two complex phasors do not fluctuate with respect to one another with time. As a result, light from two points on the object can in principle interfere, yielding a stable interference pattern. As a consequence, a fully coherent imaging system is linear in amplitude. However, when we measure the image with a detector, we can only measure intensity.

On the other hand, for a fully incoherent imaging system, the complex values of the fields at two different object points vary randomly and independently, while the intensities at various object points remain stable. The result is that the fields from those two points would not interfere, but rather would simply add on an intensity basis. As a consequence, fully incoherent optical imaging systems are linear in intensity.

7.2 The Amplitude Impulse Response and Transfer Function

As explained above, for an object illuminated by nearly monochromatic, coherent light, all object points are oscillating with fixed relative amplitudes and mutually locked phases. Since the propagation of light amplitude is a linear process (the wave equation is linear), we can express the complex distribution of light amplitude in the image plane as a superposition integral of the light complex amplitude in the object plane times an amplitude impulse response (or amplitude point-spread function) of the optical system that forms the image. Thus, if $\mathbf{U}_o(\xi, \eta)$ represents the distribution of complex amplitude across the object plane, and $\mathbf{U}_i(x, y)$ represents the distribution of complex amplitude across the image plane, then

$$\mathbf{U}_i(x, y) = \int\int_{-\infty}^{\infty} \mathbf{U}_o(\xi, \eta)\, \tilde{\mathbf{h}}(x, y; \xi, \eta)\, d\xi\, d\eta,$$

where $\tilde{\mathbf{h}}(x, y; \xi, \eta)$ is the amplitude impulse response of the optical system. In the event that the amplitude impulse response is identical for all point-source locations (ξ, η) in the object space, the superposition integral becomes a convolution:

$$\mathbf{U}_i(x, y) = \int\int_{-\infty}^{\infty} \mathbf{U}_o(\xi, \eta)\, \mathbf{h}\,(x-\xi, y-\eta)\, d\xi\, d\eta.$$

In this case, we say that the optical system is spatially invariant, or equivalently, isoplanatic.

Given the convolution relationship between object and image for an isoplanatic system, it is natural to define a corresponding mapping in the frequency domain, namely, a 2D transfer function. We shall call this transfer function the amplitude transfer function and represent it by $\mathbf{H}(\nu_x, \nu_y)$. It is defined as

$$\mathbf{H}(\nu_x, \nu_y) = \mathcal{F}\{\mathbf{h}(x, y)\},$$

where ν_x and ν_y are spatial frequencies and the symbol $\mathcal{F}\{\ \}$ indicates a 2D Fourier transform.

The forms of both the amplitude point-spread function and the amplitude transfer function depend on the geometry of the optical system. In the section that follows, we define the geometry we shall use here for illustrative purposes.

7.3 Coherent Imaging System Model

We adopt an imaging system model that is simple but has many of the properties of a variety of real imaging systems. This model is shown in the figure below:

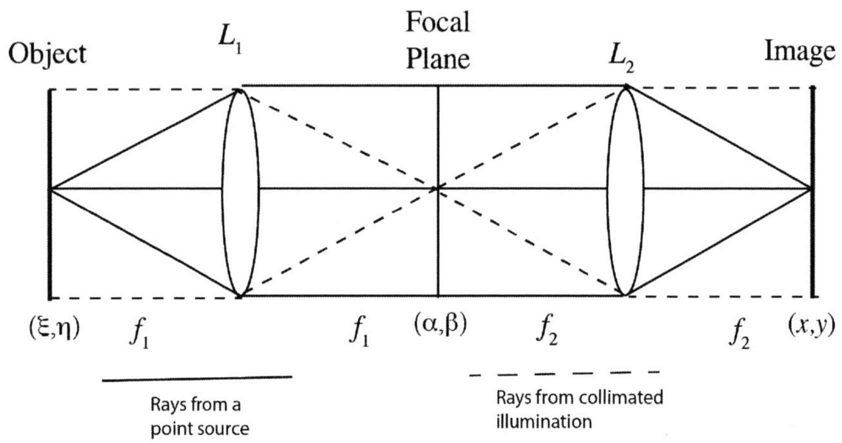

This system geometry is referred to as a $4f$ system since its total length is 4 focal lengths $(2f_1 + 2f_2)$ of the lenses. The solid lines show rays from an on-axis source point. The dotted lines show parallel rays entering and exiting the system after coming to a focus in the focal plane. Assuming that the object is illuminated by a normally incident plane wave, the focal plane contains the Fourier transform of the object amplitude transmittance and the image plane contains an inverted image of the object. Note that the coordinates in the object space are (ξ, η), those in the focal plane are (α, β), and those in the image space are (x, y).

This model is over-simplified for most real-world systems, but it does contain the fundamental properties of many cases. For example, in a microscope, L_1 is the objective lens, L_2 is the eyepiece, and the object is very close to the objective lens. In a Keplerian telescope, similar names apply to L_1 and L_2, but the object is very far from the objective lens.

Two facts to keep in mind about this and many other imaging geometries are as follows: (1) the image is an inverted version of the object; (2) if $f_2 \neq f_1$, the microscope magnifies the image in the amount $\mathbb{M} = f_2/f_1$ while the telescope magnifies the angles of arrival of rays at the objective lens by $1/\mathbb{M} = f_1/f_2$. Thus, in the microscope, the focal length of the objective lens should be much smaller than the focal length of the eyepiece, while in the telescope, the focal length of the objective lens should be much larger than the focal length of the eyepiece.

In our analyses here, for simplicity, we assume that the two focal lengths are equal, yielding unity magnification, and we ignore the inversion of the image in analytical expressions. The latter step can be justified by reversing the direction of the coordinate axes in the image plane, as compared with those directions in the object plane.

Thus, for this optical system, if we start with an object complex field distribution $\mathbf{U}_o(\xi, \eta)$, and if the amplitude transmittance through the focal plane is represented by $\mathbf{t}_A(\alpha, \beta)$, the image amplitude distribution is related to the object amplitude distribution by

$$\mathbf{U}_i(x, y) = \mathcal{F}^{-1}\{\mathbf{t}_A(\alpha, \beta) \times \mathcal{F}\{\mathbf{U}_o(\xi, \eta)\}\},$$

where $\mathcal{F}\{\}$ represents a forward Fourier transform and $\mathcal{F}^{-1}\{\}$ an inverse Fourier transform. In optics, the frequency variables are usually scaled by $1/\lambda f$, but in the simulations that follow, we ignore this scaling factor. We see that the amplitude transmittance through the focal plane will be the amplitude transfer function in our simulations.

Continuous Fourier transforms can be performed by *Mathematica®* for a certain number of functions. Therefore, we illustrate the imaging problem in the following section without discretizing any functions.

7.4 Symbolic Simulation Example

It is possible to simulate coherent imaging with the $4f$ system using either symbolic functions or discrete (sampled) functions in *Mathematica.* However, the use of the symbolic approach is much more limited than the discrete approach. It is frequently not possible to describe the object to be imaged in terms of symbolic functions, and the continuous Fourier transform algorithm (i.e., `ft2[]` for 2 dimensions in our abbreviated notation) in *Mathematica* may not know the transforms of some of the functions involved. As a result, the discrete approach is much more flexible in general, since the object can be described by a set of samples and the discrete Fourier transform (`DFT[]` in our abbreviated notation) will always perform the transform on the sampled array.

Nonetheless, to illustrate that the symbolic approach can work, we present a simple example of its use in what follows. We will then turn our attention to the discrete approach.

The Object

The object in this example is assumed to be a rectangular transparent opening with widths $\mathcal{L}_1 \times \mathcal{L}_2$ in an opaque background. The amplitude transmittance through the focal plane is taken to be a square, transparent area with dimensions $\mathcal{W} \times \mathcal{W}$. The following code calculates the intensity in the image plane. Remember that the rectangle function rect( ) equals value 1 for argument in the range ±1/2, and 0 otherwise. The *Mathematica* function `Sinc[x]` is equivalent to `sin[x]/x`.

First we specify the values of the parameters that will be assumed in the simulation.

```
In[ ]:= ℒ1 = 2; ℒ2 = 100; 𝒲 = 0.3;
```

We choose a simple object in order to be sure that *Mathematica* knows its symbolic Fourier transform.

```
In[ ]:= Uo[ξ_, η_] = rect[ξ / ℒ1] * rect[η / ℒ2];
```

Plot the amplitude distribution of this object function.

```
In[ ]:= Plot3D[Uo[ξ, η], {ξ, -50, 50}, {η, -100, 100}, PlotRange → All,
          PlotPoints → 200, Filling → Automatic, ExclusionsStyle → Automatic,
          AxesLabel → {ξ, η, I}, PlotLabel → "Object Amplitude"]
```

Out[ ]=

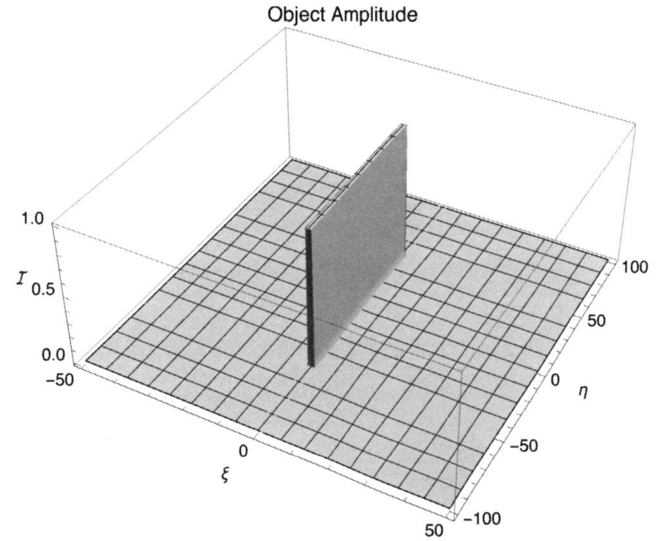

Note that the object has a long dimension in the η direction and a short dimension in the ξ direction. *Mathematica* knows the Fourier transform **F** of this object.

In[ ]:=
```
F[α_, β_] = FullSimplify[FourierTransform[Uo[ξ, η],
    {ξ, η}, {α, β}, FourierParameters → {0, -2 * Pi}]]
```

Out[ ]=
```
200 Sinc[2 π α] Sinc[100 π β]
```

Now plot the magnitude of this Fourier spectrum.

In[ ]:=
```
Plot3D[Abs[F[α, β]], {α, -1, 1}, {β, -1, 1}, PlotRange → All,
  AxesLabel → {α, β, |ℱ|}, PlotLabel → "Spectral Magnitude"]
```

Out[ ]=

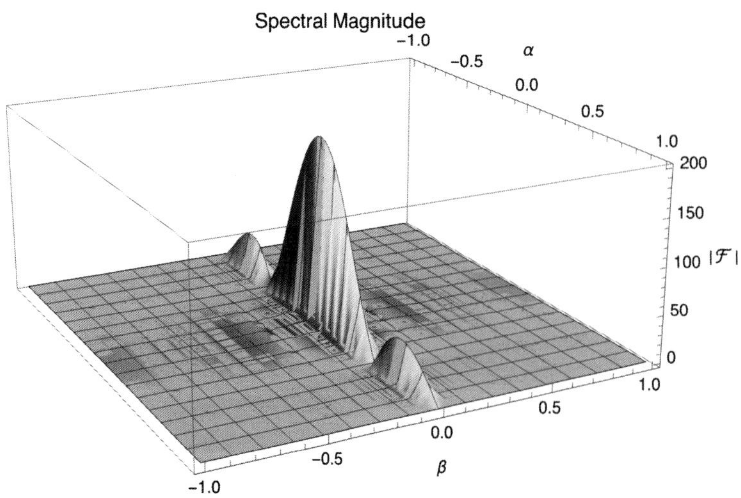

The spectrum of the object is broad in the α direction, caused by the short dimension of the object, and narrow in the β dimension, caused by the long η dimension of the object.

The Focal-Plane Filter

Now specify the amplitude transmittance through the focal plane.

```
In[ ]:= t[α_, β_] = rect[α / W] * rect[β / W];
```

Multiply the focal-plane transmittance by the object Fourier spectrum, yielding the amplitude of the field transmitted by the focal plane.

```
In[ ]:= G[α_, β_] = t[α, β] * F[α, β]
Out[ ]=
    200 Sinc[2 π α] Sinc[100 π β] UnitBox[3.33333 α] UnitBox[3.33333 β]
```

Remember that *Mathematica*'s `UnitBox[arg]` is identical with the user-defined function `rect[arg]`, and `Sinc[arg]` is equivalent to the user-defined function `sinc[arg/π]`.

Plot the magnitude of this filtered Fourier spectrum.

```
In[ ]:= Plot3D[Abs[G[α, β]], {α, -1, 1}, {β, -1, 1}, PlotRange → All,
     ExclusionsStyle → Automatic, AxesLabel → {α, β, |G|},
     PlotLabel → "Filtered Spectral Magnitude"]
```

Out[ ]=

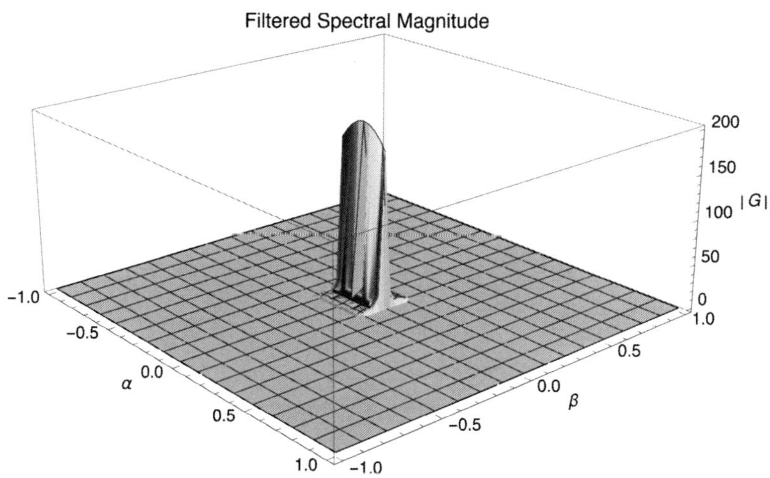

The cutoff frequency is 0.3 in units that are the reciprocal of the units chosen for ξ and η.

The Image

Find the image complex field by inverse Fourier transforming the field transmitted through the focal plane. Because we defined the shorthand inverse Fourier transform `ifft2[]` with different variables, we use the full *Mathematica* form of the inverse transform here.

```
In[ ]:= Uᵢ[x_, y_] = InverseFourierTransform[G[α, β],
          {α, β}, {x, y}, FourierParameters → {1, -2 * Pi}]
Out[ ]=
 1
 ── (0.398942 SinIntegral[0.942478 - 0.942478 x] +
 π²
     0.398942 SinIntegral[0.942478 + 0.942478 x])
   (0.398942 SinIntegral[47.1239 - 0.942478 y] +
     0.398942 SinIntegral[47.1239 + 0.942478 y])
```

This is a complicated result, but the function **SinIntegral[]** is known to *Mathematica*. Plot the intensity in the image plane.

```
In[ ]:= I[x_, y_] = Abs[Uᵢ[x, y]]^2;
        Plot3D[I[x, y], {x, -50, 50},
         {y, -100, 100}, PlotRange → All, PlotPoints → 200,
         AxesLabel → {x, y, I}, PlotLabel → "Image Intensity"]
Out[ ]=
```

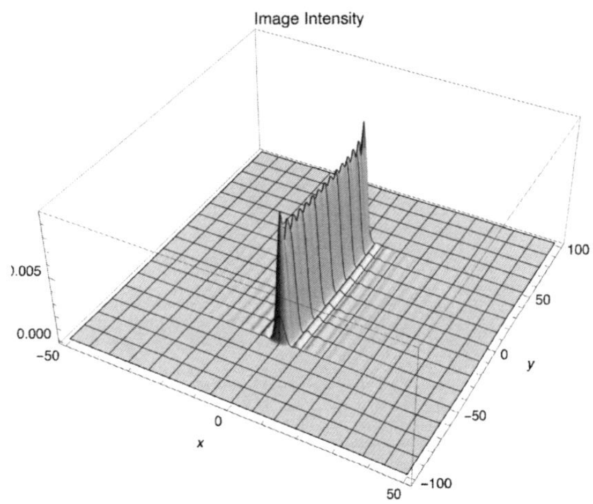

We see that *Mathematica* has been able to perform all of these computations in a symbolic fashion (except for plotting, which always involves sampling). This has been possible because of our choice of a simple object function and a simple amplitude transmittance in the focal plane. *Mathematica* does know a vast set of symbolic Fourier transforms, but it is not difficult to find examples for which *Mathematica* does not know the Fourier transform of the object function or the inverse Fourier transform of the filtered spectrum, and therefore the computation is halted. This is particularly true for truncated functions, even though the Fourier transform of the untruncated transform may be known. An example of the untruncated function follows.

```
In[ ]:= ft2[1 / Sqrt[ξ^2 + η^2]]
```

$$\frac{1}{\sqrt{\nu x^2 + \nu y^2}}$$

Note that this function is its own transform. *Mathematica* knows this symbolic Fourier transform. Now limit the same function with a truncation in the (ξ, η) domain.

```
In[•]:= ft2[1 / Sqrt[ξ^2 + η^2] * rect[ξ] * rect[η]]
```

$$\text{FourierTransform}\left[\frac{\text{UnitBox}[\eta]\ \text{UnitBox}[\xi]}{\sqrt{\eta^2 + \xi^2}},\right.$$
$$\left.\{\xi, \eta\}, \{\nu x, \nu y\}, \text{FourierParameters} \to \{0, -2\pi\}\right]$$

Mathematica cannot perform this transform. Because of these limitations of the symbolic approach, we turn to a discrete approach, which in general is much more flexible.

7.5 Discrete Imaging Examples

The previous example was based on symbolic functions and symbolic transforms. We turn now to the case of a discrete object and discrete transforms.

An Amplitude Object

We choose as our object for this simulation a standard U.S. Air Force test target, as shown below. The target has three color channels (red, green, and blue) and is in *Mathematica*'s Image format in its original form.

We first convert the object to a grayscale image.

```
In[•]:= testtarget = ColorConvert[
```

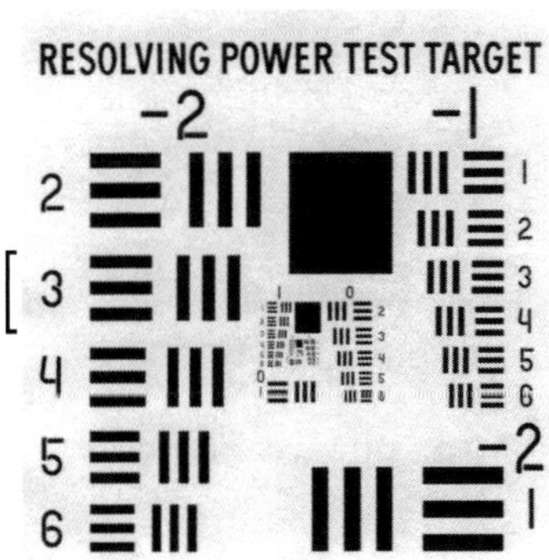


```
, "Grayscale"];
```

Find the number of pixels in this image.

```
In[•]:= ImageDimensions[testtarget]

{3148, 3250}
```

This image consists of more than ten million discrete pixels. We reduce the target to a 1024 × 1024 array of pixels. To perform this operation, we must first convert from a *Mathematica* Image format to a Data format (the command **ImageData[]** performs this conversion), and then downsample the image by factors of 3 in each dimension.

```
In[•]:= newdata = Downsample[ImageData[testtarget], {3, 3}];
Dimensions[%]

{1084, 1050}
```

The number of pixels has not yet been converted to 1024 × 1024.

Using the **Take[]** command, we reduce the number of data samples to the desired number. The conversion from Image to Data format has reversed the order of the pixels in both dimensions. This can be countered with a **Reverse[]** command.

```
In[ ]:= object = Reverse[Take[newdata, {40, 1063}, {13, 1036}]];
In[ ]:= Dimensions[object]
        {1024, 1024}
```

The pixels are now in the proper order, as the density plot below shows.

```
In[ ]:= ListDensityPlot[object, ColorFunction → GrayLevel, ImageSize → 300]
Out[ ]=
```

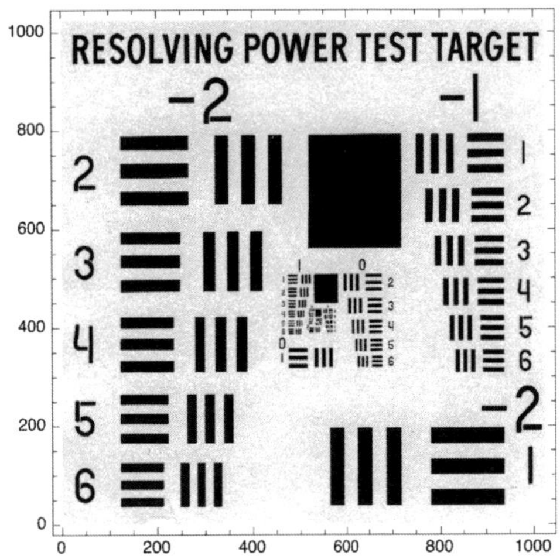

We will use this reduced image in later chapters as well as here, so we export it to a data file. We can input this data file wherever needed with the command **Import["filename"]**.

```
In[ ]:= Export["Barchartdata.dat", object];
```

A Focal-Plane Amplitude Filter

The next task is to specify a focal-plane filter (a "stop," or a filter that passes certain Fourier components and blocks others) to be applied to the object. We choose to use a circular passband filter with diameter $\mathcal{D}$ in the focal plane.

```
In[ ]:= N = 1024; D = 128;
In[ ]:= stop = Parallelize[
          Table[rect[Sqrt[(α - N / 2) ^2 + (β - N / 2) ^2] / (D / 2)  - 1 / 2],
           {α, 1, N}, {β, 1, N}]];
In[ ]:= ListPlot3D[stop, AxesLabel → {α, β, "Transmittance"},
         PlotLabel → "Fourier-Plane Filter"]
```

Out[ ]=

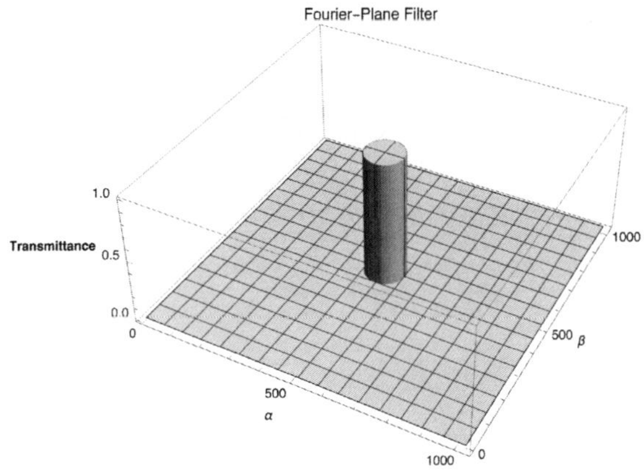

It remains to apply this filter to the Fourier spectrum of the object and display the resulting intensity distribution for the image.

The Image Intensity

To find the complex amplitude distribution in the image plane, we must Fourier transform the object, multiply the resulting spectrum by the filter transfer function, and inverse Fourier transform the result. The image intensity is then the squared magnitude of the image complex amplitude.

```
In[ ]:= imageampl = idft[RotateRight[Fourier[object], {N / 2, N / 2}] * stop];

In[ ]:= Image[10^6 * Reverse[Abs[imageampl]^2, {2, 2}], ImageSize → 300]
```

Out[ ]=

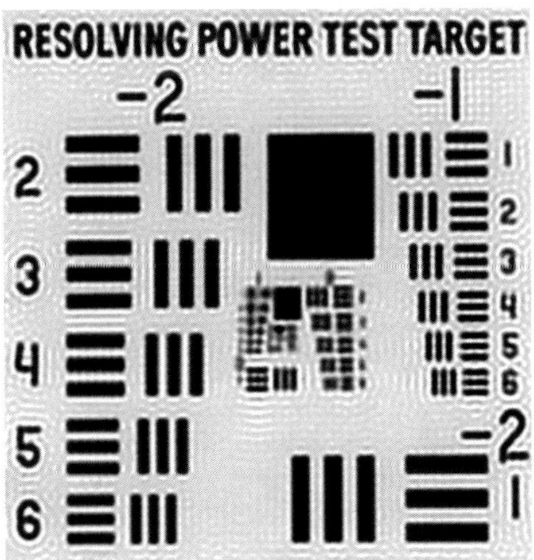

The small bar groups under the large black box have been mostly blurred to the point that they cannot be discerned. Finally, the large bar groups exhibit ringing caused by the sharp cutoff of the Fourier-plane filter.

A Phase Object and a Focal-Plane Phase Filter

It is also possible to construct filters in the focal plane that consist entirely of phase changes. An excellent example arises in the case of Zernike phase contrast imaging, by means of which we can image a pure phase object by converting phase to intensity. To demonstrate this case, we first create an object consisting solely of phase shifts. We accomplish this by creating an object of the form

$$\mathbf{U}_o(\xi, \eta) = e^{i\,\theta(\xi,\eta)},$$

where θ is the structure of the phase of the phase-only object we wish to visualize.

We first define the phase profile of the object we wish to image, and show it in a 3D plot.

```
In[ ]:= N = 1024;
   phaseprofile =
   Parallelize[Table[tri[Sqrt[((ξ - N / 8) ^2 + (η - N / 8) ^2)] / 16] +
      tri[Sqrt[((ξ + N / 8) ^2 + (η + N / 8) ^2)] / 64] +
      tri[Sqrt[((ξ + N / 4) ^2 + (η - N / 4) ^2)] / 32],
     {ξ, -N / 2 + 1, N / 2}, {η, -N / 2 + 1, N / 2}]];
  ListPlot3D[phaseprofile,
   PlotRange → All, AxesLabel → {"ξ", "η", "Phase"}]
Out[ ]=
```

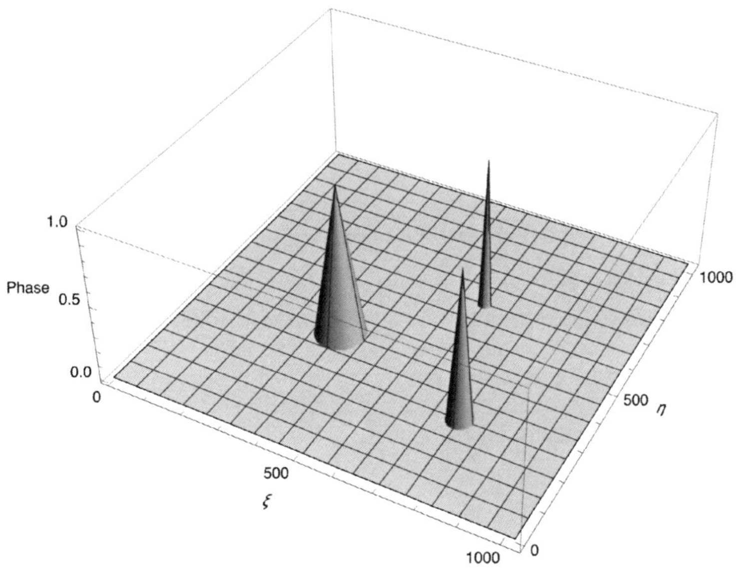

Now define the phase object using the phase profile just defined. We choose the maximum phase change to be $\pi/8$, and we plot the intensity of the image of the pure phase object.

```
In[ ]:= phaseobject = Exp[I * (Pi / 8) * phaseprofile];
   ListPlot3D[Abs[phaseobject] ^2,
    ImageSize → 350, AxesLabel → {"ξ", "η", "Intensity"}]
Out[ ]=
```

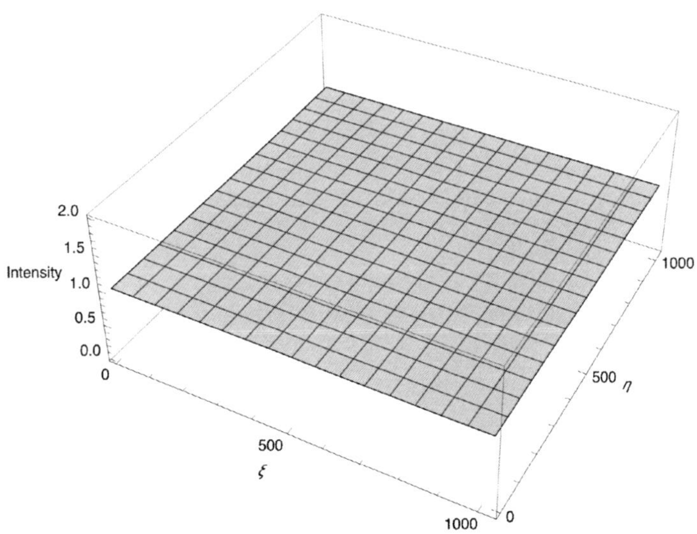

The phase structures on the object are not visible. The entire image intensity has a value of unity.

The next step is to Fourier transform the phase object, thus finding its Fourier spectrum.

```
In[*]:= spectrum = RotateRight[dft[phaseobject], {N / 2 , N / 2 }];
```

The Phase-Only Filter and the Image

Our attention is now turned to the Zernike phase-shift dot that must be placed at the origin in the Fourier domain. To allow experimentation with the effects of different phase-shifting dot sizes, we make it circular with a general radius r that can be changed by the reader with access to the *Mathematica* notebook.

```
In[*]:= psdot[r_] = Exp[I * (Pi / 4) * Parallelize[
          Table[circ[Sqrt[((α - N / 2 - 1) ^2 + (β - N / 2 - 1) ^2)] / r],
           {α, 1, N}, {β, 1, N}]]];
```

If we choose the radius of this dot to be 1, we obtain the following image.

```
In[*]:= image = Abs[idft[spectrum * psdot[1]]] ^2;
```

```
In[*]:= ListPlot3D[image, PlotRange → All,
         ImageSize → 350, AxesLabel → {"x", "y", "Intensity"}]
```

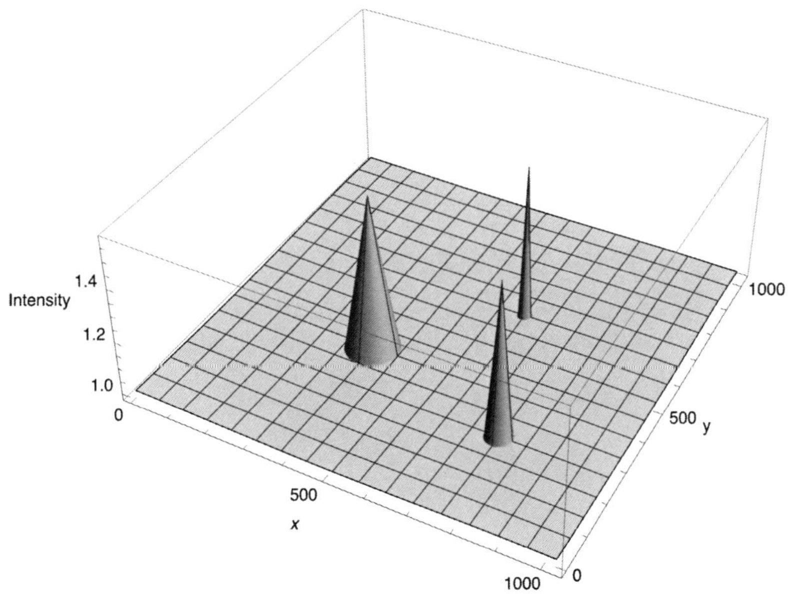

Note that the images of the phase objects ride on a background of value 1, the same as the background of the original phase-only object. The three peaks appear clearly in the intensity image, even though they were not visible in the intensity of the original phase object.

If the phase-shifting dot in the focal plane is too wide (radius 10 in the next example), the phase structures begin to disappear into the intensity background.

```
In[*]:= image2 = Abs[idft[spectrum * psdot[10]]] ^2;
        ListPlot3D[image2, PlotRange → All,
         ImageSize → 350, AxesLabel → {"x", "y", "Intensity"}]
```

Out[ ]=

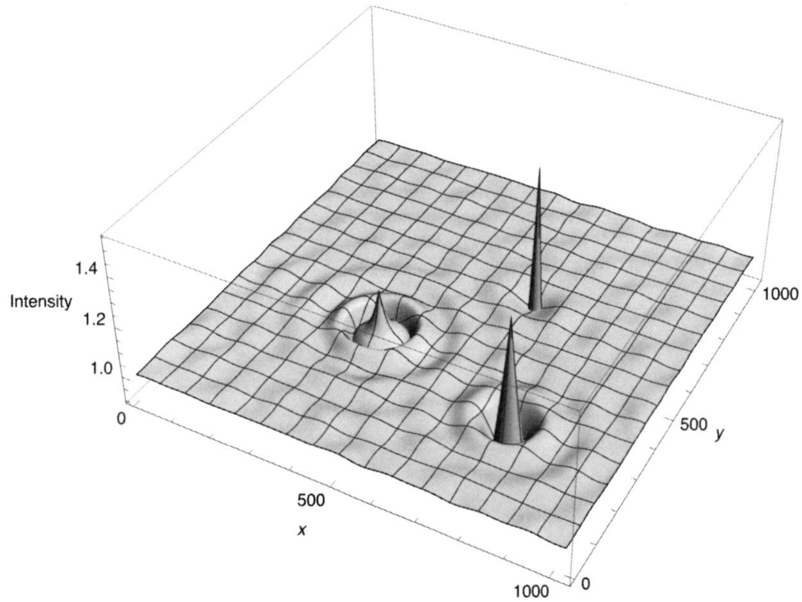

The larger pyramids disappear first because they have the narrowest spectra and are most affected by the overly large phase-shifting dot.

7.6 Coherent Image Speckle

The above simulations have neglected the fact that in many coherent imaging problems, the object of interest is rough on the scale of a wavelength, or alternatively has diffuse transmittance. To account for this object characteristic, it is necessary to explicitly introduce a random phase structure (i.e., a diffuser) in the object plane, sandwiched with the information-bearing object.

The first step in simulating this effect is to create a model for the diffuser. To do so, we start with a model for the random phase fluctuations, modeling each phase sample at each point as an independent random variable with a uniform probability density over the range $(0, 2\pi)$, implicitly assuming that the rough surface (or diffuser) contains random optical path excursions of at least several wavelengths.

```
In[ ]:= N = 1024;
        phase = RandomReal[{0, 2 * π}, {N, N}];
```

The amplitude transmittance of the diffuser is then

```
In[ ]:= diffuser = Exp[I * phase];
```

The total transmittance through the object plane is

```
In[ ]:= newobject = diffuser * object;
```

From this point on, the simulation follows that of the previous subsection.

```
In[ ]:= imageampl2 = Reverse[idft[
            RotateRight[Fourier[newobject], {N / 2, N / 2}] * stop], {2, 2}];

In[ ]:= Image[4 * Abs[imageampl2] ^ 2 / Max[Abs[imageampl2] ^ 2],
         ImageSize → 300]
```

Out[ ]=

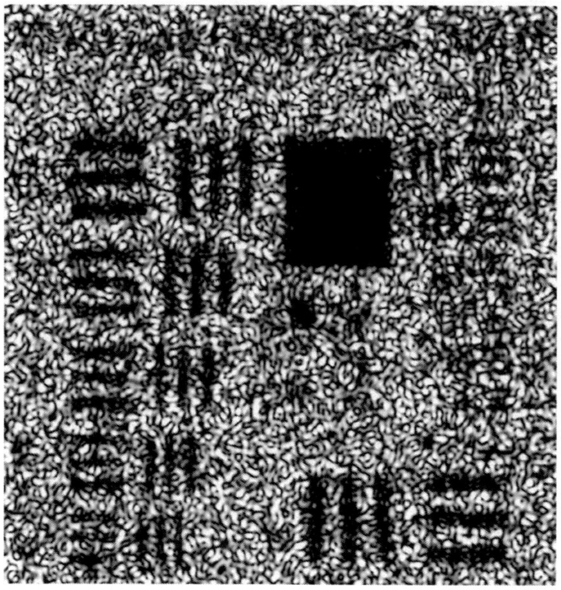

The factor of 4 used above is there simply to adjust the brightness of the image. The **Reverse[]** command is required for proper image orientation when converting data to an image.

Note that the same stop has been used here as was used when the diffuser was absent. The random granular structure in the image intensity is known as "speckle" and yields a further degradation of the image, reducing resolution even further than that observed before. For much more detailed discussions of simulating speckle, see Refs. [9] and [10].

We have pointed out a number of undesirable properties of coherent images. Some imaging techniques, such as holography and more generally interferometry, require coherent light. However, many other problems allow for the use of incoherent light, so we focus the next chapter on incoherent imaging.

8. Imaging With Incoherent Light

In this chapter, we simulate the properties of imaging systems that use incoherent light. While coherent light generally arises from a narrowband optical source that has very small spatial extent (such as a laser), incoherent light arises when the source is spatially extended, with all source points radiating independently. While coherent imaging systems are linear in complex field amplitude, incoherent systems are linear in intensity. For additional information on incoherent systems, see Ref. [4], Section 7.3.

8.1 The Intensity Impulse Response and the Optical Transfer Function (OTF)

We again assume the $4f$ system of the previous chapter, again with object coordinates (ξ, η), focal plane coordinates (α, β), and image plane coordinates (x, y). This time the light illuminating the object plane is assumed to be incoherent, with each source point radiating independently. For a discussion regarding criteria under which an imaging system is fully incoherent, see Ref. [6], p. 283.

If we again assume isoplanicity (i.e., spatial invariance), the relation between object and image becomes

$$\mathcal{I}_i(x, y) = \iint_{-\infty}^{\infty} \mathcal{I}_o(\xi, \eta)\, |\mathbf{h}(x-\xi, y-\eta)|^2\, d\xi\, d\eta,$$

where $\mathcal{I}_o$ and $\mathcal{I}_i$ represent the object and image intensity distributions, respectively, and $\mathbf{h}$ is the same amplitude impulse response encountered in the coherent imaging case, but subjected to a magnitude-squared operation to form the intensity impulse response. We again see a convolution relation, but this time the (intensity) impulse response is given by $|\mathbf{h}(x, y)|^2$.

Whenever a convolution appears, our instincts should take us to the frequency plane, where a transfer function can be defined. The transfer function for incoherent imaging is called the optical transfer function, abbreviated OTF, and is the normalized Fourier transform of the intensity impulse response. The normalization constant is by custom the value of the transfer function at zero frequency. In addition, since the OTF is found by Fourier transforming $|\mathbf{h}(x, y)|^2$, the autocorrelation theorem of Fourier analysis implies that, for the $4f$ imaging system, the OTF is a normalized autocorrelation function of the amplitude transfer function. In turn, the amplitude transfer function is a scaled version of the amplitude transmittance $\mathbf{t}_A$ through the focal plane, where again in simulations, we ignore the scaling factor $1/\lambda f$ and the frequency variables are taken to be the same as the space variables (α, β) in the focal plane.

We conclude that the OTF $\mathcal{H}$ is given in terms of the amplitude transfer function $\mathbf{H}$ by

$$\mathcal{H}(\nu_X, \nu_Y) = \frac{\iint_{-\infty}^{\infty} \mathbf{H}(\alpha, \beta)\, \mathbf{H}^*(\alpha-\nu_X, \beta-\nu_Y)\, d\alpha\, d\beta}{\iint_{-\infty}^{\infty} \mathbf{H}(\alpha, \beta)\, \mathbf{H}^*(\alpha, \beta)\, d\alpha\, d\beta},$$

and, for the simulations that follow, we take $\mathbf{H}$ to equal the amplitude transmittance through the focal plane, $\mathbf{t}_A(\alpha, \beta)$.

8.2 The Object

We assume the same 1024×1024 grayscale object used in the previous chapter, i.e., the U.S. Air Force test target, retrieving it from the file that was saved. This time, since we are imaging the intensity distribution across the object, we apply a magnitude-squared operation to the object. Since the object is largely black and white, it doesn't change much.

```
In[•]:= dataobject = Import["Barchartdata.dat"];
```

```
In[•]:= Dimensions[dataobject]
```

```
{1024, 1024}
```

```
In[•]:= intensityobject = Abs[dataobject]^2;
        Image[Reverse[intensityobject], ImageSize → 300]
```

Out[•]=

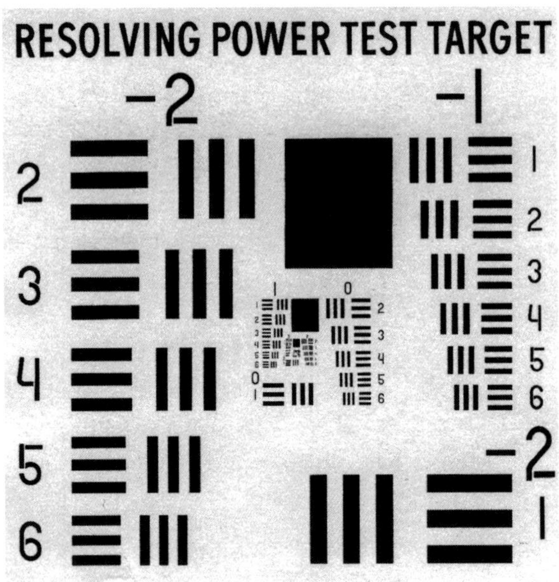

For this and future use, note that there are two different methods that could be used to display an object or the image of an object. We could use either of the following methods:

```
ListDensityPlot [dataarray]
ImageAdjust [Image [Reverse [dataarray ]]]
```

The second of these methods is much faster than the first. It was used for the image above and will be used for most of the images that follow, both in this chapter and others. The **Reverse[]** command is needed with the **Image[]** command to undo the reversal that occurs with this method. The **ImageAdjust[]** command brings the values of the image array into a range that is expected by the **Image[]** command.

8.3 The Focal-Plane Aperture

Again we assume that the transmittance through the focal plane is a circular opening in an opaque background, the diameter of the opening being represented by $\mathcal{D}$. We take the same values for $\mathcal{N}$ and $\mathcal{D}$ used in the previous chapter and define the focal-plane stop as before:

```
In[•]:= N = 1024; D = 128;
        stop = Parallelize[
           Table[rect[Sqrt[(α - N / 2)^2 + (β - N / 2)^2] / (D / 2) - 1 / 2],
           {α, 1, N}, {β, 1, N}]];
```

Next we calculate the OTF by auto-correlating the amplitude transmittance $t_A(\alpha, \beta)$ = stop through the focal plane.

8.4 The Optical Transfer Function

To find the optical transfer function, we must find the autocorrelation function of the data array **stop**. This can be accomplished by finding the Fourier transform of **stop**, taking its squared magnitude, and inverse Fourier transforming the result.

```
In[•]:= FT = dft[stop];
        sqmagnitude = Abs[FT]^2;
        unnormalized = RotateRight[idft[sqmagnitude], {N/2, N/2}];
        OTF = Abs[unnormalized] / Abs[unnormalized][[N/2, N/2]];
```

```
In[•]:= ListPlot3D[Abs[OTF], PlotRange → All, AxesLabel → {νX, νY}]
```

Out[•]=

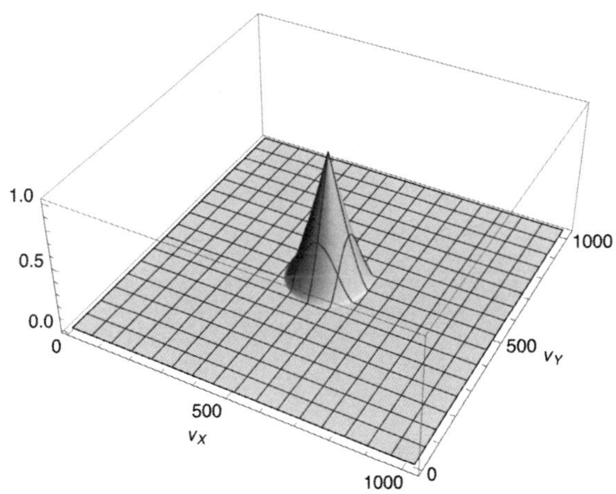

Note that an absolute value is applied in the **ListPlot** command because the result of Fourier transforming yields some very small imaginary parts. Plotting just the real part of the OTF yields the same result as a plot of the magnitude of the OTF.

Because of the autocorrelation operation applied to the amplitude transfer function, the cutoff frequency (i.e., the frequency where the transfer function goes to zero) for incoherent imaging is twice the cutoff frequency for coherent imaging. However, you should not place much significance on this fact, since in the coherent system we are dealing with amplitude while in the incoherent system we are dealing with intensity. As a result, a comparison of achievable resolution is more complex than just this comparison of cutoff frequencies.

8.5 Finding the Incoherent Image

To find the image, we must now multiply the OTF by the Fourier transform of the object and inverse Fourier transform the result. The following code accomplishes this:

```
In[•]:= image = idft[OTF * RotateRight[dft[dataobject], {N/2, N/2}]];
```

```
In[•]:= Image[Reverse[Abs[image]], ImageSize → 300]
```

Out[ ]=

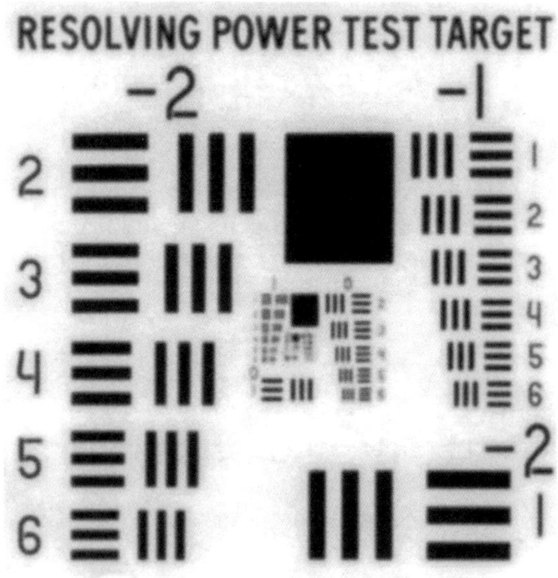

The incoherent image differs from the coherent image in several different aspects. First, the background level of the image is brighter than that of the coherent image, presumably because the OTF passes more Fourier components of the object than in the coherent case. Second, there is no ringing at the edges of the bars, due to the absence of a sharp cutoff of the OTF. Third, we can resolve most of the bar triplets in the group labeled 0 under the large dark square, something we could not do in the coherent case. Lastly, even if a diffuser is placed in contact with the object, there will be no speckle observed in the image, since we are imaging the magnitude squared (intensity) of the object, for which phase does not matter.

By comparing the images obtained in the previous chapter and this chapter, we see that there are many advantages to imaging with incoherent light rather than coherent light. Why then is coherent imaging of interest? Primarily because there are some imaging modalities that require coherent light (e.g., holography) and because some imaging systems are close to coherent in behavior, even though the illumination may be from an incoherent source. See Ref. [6], p. 283, for elaboration on this latter point.

9. Imaging With Holography

Holography is a coherent imaging method, but quite different than the method using the $4f$ system, so we devote a chapter to this subject. There many different methods for recording holograms and retrieving images from them. See Ref. [4] for a more detailed discussion than presented here. In this chapter we simulate only a few of the most common methods.

9.1 Gabor Holography

Holography was invented by Dennis Gabor in 1948 (Ref. [11]). Our first simulation will be of Gabor's method for recording holograms and viewing the images they produce. Gabor's method depends on having an object to be imaged that is largely transparent, perhaps with small, darker structures we wish to visualize.

The hologram is formed by interferometry. In this case, a "reference" wave is supplied by the coherent light transmitted by the transparent part of the object, and an "object" wave results from diffraction and scattering of coherent light by the smaller objects to be visualized. The interference pattern intensity that results is recorded at distance z from the object to form the hologram, as shown below.

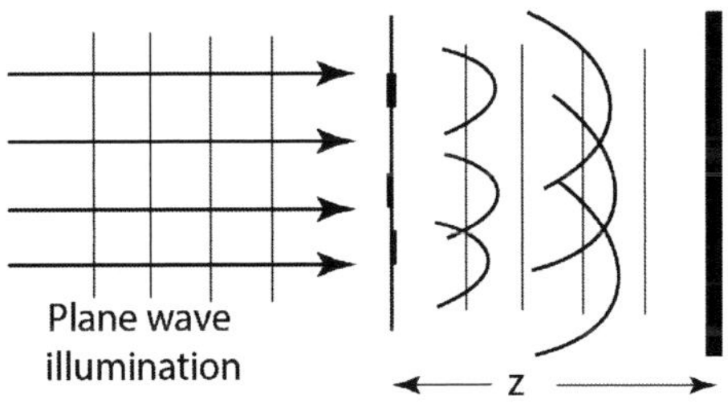

Mathematics of the Gabor Hologram

The mathematics behind the Gabor hologram is quite simple. Let $\mathcal{R}$ represent the reference wave arriving at the recording medium and let $\mathcal{O}$ represent the light scattered and diffracted by the object structures of interest. The reference $\mathcal{R}$ is a plane wave in this case. In the case of the Gabor hologram, both $\mathcal{R}$ and $\mathcal{O}$ arise from the object transparency. When $\mathcal{R}$ and $\mathcal{O}$ interfere at the hologram, the intensity $\mathcal{I}$, which is the measurable quantity, is given by

$$\mathcal{I} = |\mathcal{R} + \mathcal{O}|^2 = |\mathcal{R}|^2 + |\mathcal{O}|^2 + \mathcal{R}^* \mathcal{O} + \mathcal{R}\mathcal{O}^*.$$

The first two terms are intensity biases. The third term leads to an image in the original location of the object. This image is often called the "virtual" image and can be seen looking into the hologram from its right side. The fourth term leads to an image at distance z to the right of the hologram (the "real" image).

Recording the Hologram and Reconstructing Images

In Gabor's time, photographic film was the recording medium, and images were formed by illuminating the developed film transparency with coherent light. To achieve high-quality images it was necessary for the film to be developed in such a way that its amplitude transmittance is proportional to the intensity present during recording. The holographic images could then be viewed by illuminating the film transparency with a plane wave and looking in planes at distances $\pm z$ from the hologram.

Modern-day holography seldom uses film for recording the hologram because electronic detectors now have the resolution and sensitivity to replace silver-halide technology in most applications. In addition, in many cases, optical reconstruction of images has given way to digital reconstruction, in which the propagation of light from the recorded intensity pattern is performed digitally, treating the intensity recorded in the hologram as complex amplitude to be propagated to the image. The real image can be obtained by propagating the light transmitted by the hologram a distance $+z$ to the right. A virtual image can be obtained from the hologram by digitally back-propagating the light to the left of the hologram a distance $-z$.

We are able here to simulate both parts of the image-forming process, calculating the intensity that would fall on the detector for a given object and distance z, and then propagating that intensity pattern, treated as a complex-amplitude pattern, an appropriate distance to the image.

The Object

The **testtarget** object we have used in previous chapters generally satisfies the requirements of the Gabor hologram, namely, that it consists of a transparent background on which there are scattered opaque structures. We retrieve the data that describe this object from a stored file.

```
In[ ]:= dataobject1 = Import["Barchartdata.dat"];
        Image[Reverse[dataobject1]]
```

Out[ ]=

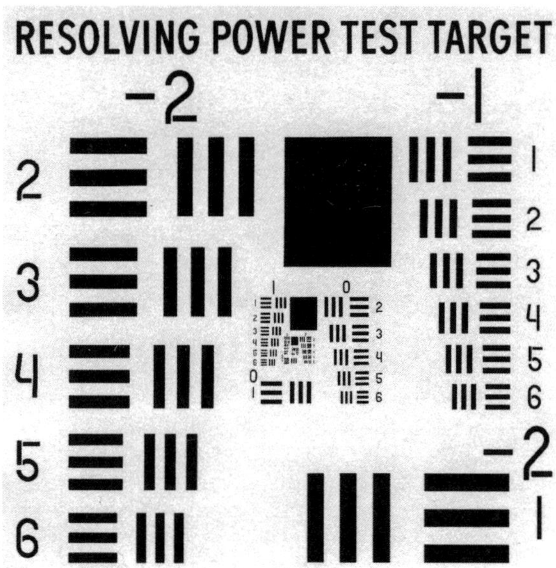


```
In[ ]:= n = Length[dataobject1]

        1024
```

A Discussion of N_F and Q

The Fresnel number is a concept that is most meaningfully applied to circular or square apertures with no internal structure. Here we are using an object bounded by a square aperture but with internal structure.

We use the Fresnel transfer function method for propagating the light, where the transfer function is given by [Ref. 4, p. 137]:

$$\mathcal{H}(k, p) = \exp\left[-i\pi / \left(4 Q^2 N_F\right) \times \left((k - N/2 - 1)^2 + (p - N/2 - 1)^2\right)\right],$$

where k and p range from 1 to N in steps of unity. As can be seen, it is the product of Q^2 and N_F, as well as the sample spacing, that determine whether there is aliasing of the transfer function. The sample spacing is fixed, so it is Q^2 and N_F that must be chosen large enough to assure insignificant aliasing. The meaning of N_F can be retained as the Fresnel number of the finite aperture, with aliasing being determined by the choice of Q^2. The usual interpretation of Q as the ratio N/M, where N is the linear dimension of the simulation and M is the number of samples within the object, is abandoned here. The required value of Q is found by trial and error.

Propagating the Light to the Recording Plane

The first step in simulating the Gabor hologram is to find the spectrum of the object.

```
In[ ]:= spectrum1 = RotateRight[dft[dataobject1], {n / 2, n / 2}];
        Image[0.002 * Reverse[Abs[spectrum1]]]
```

Out[ ]=

As can be seen, the spectrum is confined to a central region about the origin, except for spectral components that extend along the horizontal and vertical axes caused by the internal structure of the test target. Next we calculate the transfer function of propagation, choosing values for N_F and Q.

```
In[ ]:= q = 10; nf = 4;
        ℋ = Parallelize[Table[
            Exp[-I * π * (1 / (4 * q^2 * nf)) * ((k - n / 2 - 1) ^2 + (p - n / 2 - 1) ^2)],
            {k, 1, n}, {p, 1, n}]];
        ImageAdjust[Image[Reverse[Re[ℋ]]]]
```

Out[ ]=

The quantity $\mathcal{H}$ represents the transfer function of free-space propagation. If the complex exponential argument has a negative sign, propagation is to the right of the object. Now we multiply the complex spectrum of the object by the complex transfer function, propagate to the right of the object, and inverse Fourier transform to find the complex field in the hologram plane. The intensity in the hologram plane is then calculated.

```
In[◦]:= amplitude = idft[spectrum1 * ℋ];
        intensity = Abs[amplitude] ^2;
        Image[Reverse[intensity]]
```

Out[◦]=

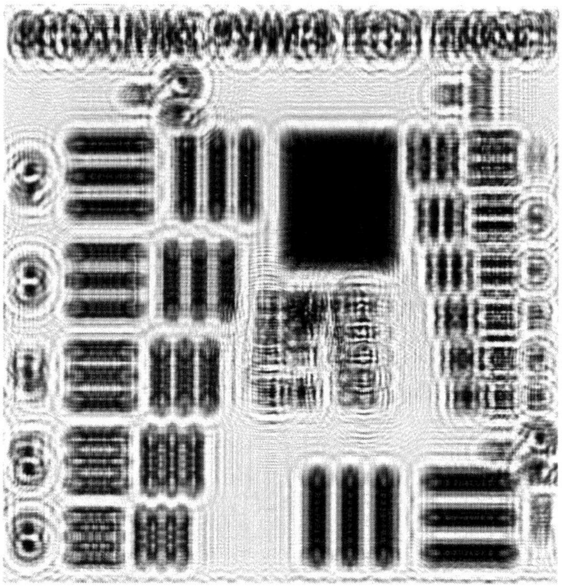

As can be seen, propagation away from the object to the hologram plane has seriously blurred the original object. Now we show that an image can be obtained from this hologram.

Reconstructing the Real Image From the Hologram

We treat the intensity detected in the hologram as a complex field to be propagated to the right, hopefully yielding a real image of the object. Again, the sign of the argument of the complex exponential in the transfer function is negative, meaning propagation to the right.

```
In[◦]:= holoamplitude = intensity;
        spectrum2 = RotateRight[dft[holoamplitude], {(n / 2), n / 2}];
        ℋ = Parallelize[Table[
            Exp[-I * π * (1 / (4 * q^2 * nf)) * ((k - n / 2 - 1) ^2 + (p - n / 2 - 1) ^2)],
            {k, 1, n}, {p, 1, n}]];

In[◦]:= imageampl = idft[spectrum2 * ℋ];
        imageint = Abs[imageampl] ^2;

In[◦]:= Image[Reverse[imageint]]
```

Out[ ]=

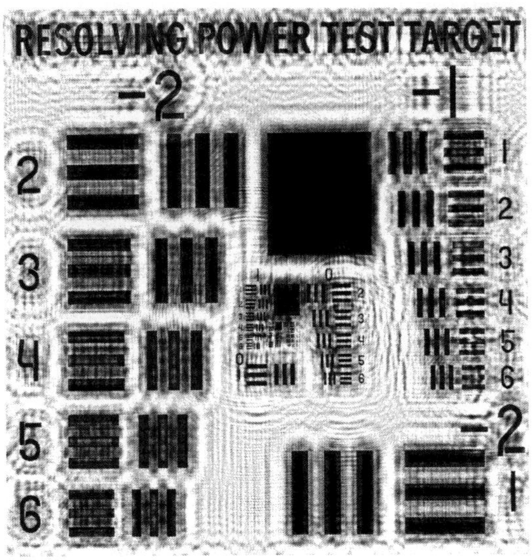

Note that sharp numerals and bars can be seen on top of background clutter. The real image has been retrieved, but it is accompanied by background noise. Part of this noise is the de-focused virtual image and part is from the on-axis terms of the hologram transmission. With the Gabor hologram, there are twin images, real and virtual, that are generated and overlap one another in the reconstructed image plane. The "twin image" problem and the overlapping presence of the light from on-axis terms of transmittance are the most important disadvantages of the Gabor hologram.

Reconstructing the Virtual Image From the Hologram

To reconstruct the virtual image, we must digitally back-propagate the field transmitted by the hologram to the original location of the object. This requires simply conjugating the transfer function, and then proceeding as before.

```
In[ ]:= ℋ = Parallelize[Table[
        Exp[-I*π*(1/(4*q^2*nf))*((k - n/2 - 1)^2 + (p - n/2 - 1)^2)],
        {k, 1, n}, {p, 1, n}]];
    imageampl = idft[spectrum2*Conjugate[ℋ]];
    imageint = Abs[imageampl]^2;
    Image[Reverse[imageint]]
```

Out[ ]=

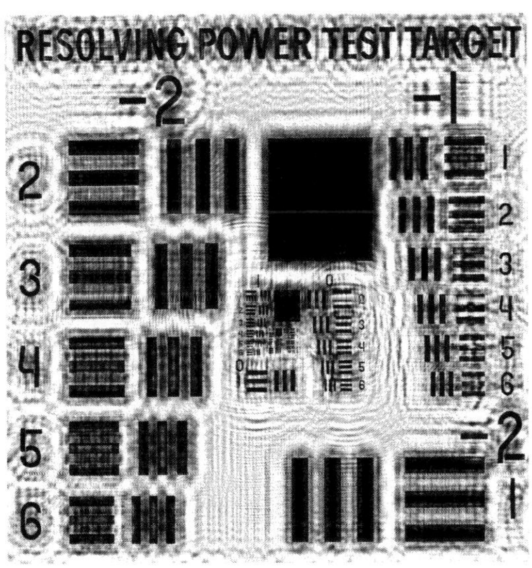

This image is identical with the previous image. Given that the background noise associated with the images reconstructed from Gabor holograms is so severe, we turn to Leith–Upatnieks holograms, which solve the twin-image noise problem.

9.2 Leith–Upatnieks Holography – The Fourier Hologram

In 1962, Leith and Upatnieks [12] introduced the idea of the offset-reference hologram, in which rather than depending on the object to supply the reference through its level of transparency, the reference wave is supplied separately from the same source as the object illumination. The original demonstration of this idea was performed with non-laser light, but the applicability was greatly strengthened by the subsequent availability of highly coherent laser light.

There are many forms of the Leith–Upatnieks hologram, but below we illustrate the simplest of those to simulate, namely the *Fourier-transform hologram*.

The Fourier-Transform Hologram Recording and Reconstruction

The recording and reconstruction geometries for this type of hologram are illustrated below.

In[ ]:=

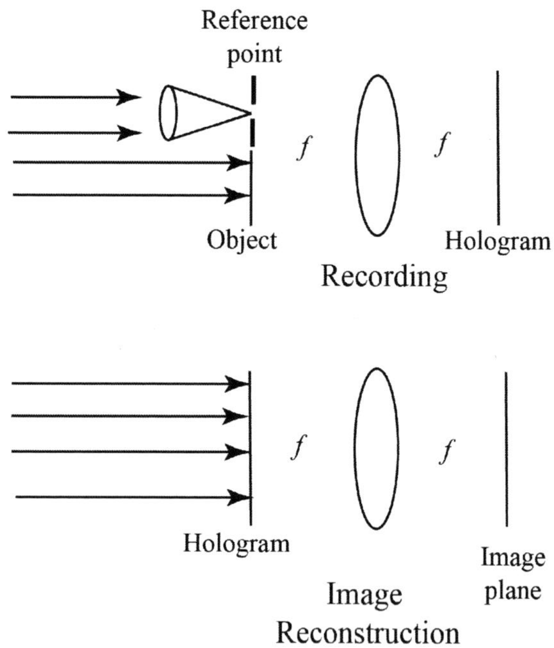

In the recording geometry, a portion of the illumination is focused to a point in the plane of the object, but separated from the object by some distance. The object and hologram planes both lie in the focal planes of the large lens. This focused point generates a reference plane wave at the hologram recording plane, arriving at an angle. The object is illuminated, and the Fourier transform of its amplitude transmittance is incident on the hologram plane as an amplitude distribution, where it interferes with the off-axis reference wave. Finally, the total incident intensity at the hologram plane is recorded.

In the image reconstruction geometry, the hologram is assumed to have an amplitude transmittance that is proportional to the intensity incident during recording. The transmitted amplitude is Fourier transformed, and the squared magnitude of that incident amplitude is the intensity distribution across the image plane.

Note that the combined recording and construction geometry can be regarded as a $4f$ system with a nonlinearity in the hologram plane provided by the transformation of incident intensity into transmitted amplitude.

Fourier-Transform Hologram Mathematical Description

The basic hologram formula introduced in the previous section again holds, and we repeat it here with a small change. The intensity incident on the hologram is given by

$$\mathcal{I} = \left|\mathcal{R} + \hat{O}\right|^2 = |\mathcal{R}|^2 + |\hat{O}|^2 + \mathcal{R}^* \hat{O} + \mathcal{R}\hat{O}^*,$$

where for this geometry, $\hat{O}$ is the Fourier transform of the amplitude transmittance of the object while $\mathcal{R}$ is a plane-wave reference, $\mathcal{R} = \exp(-i2\pi\alpha)$, with $\alpha = \sin(\theta)/\lambda$, and θ being the angle of arrival of the reference wave with respect to the z axis. This intensity distribution becomes the amplitude transmittance of the hologram in the reconstruction step.

The off-axis reference wave is the key to separating the twin images that would be present in a Gabor hologram.

The inverse Fourier transform of the $|\mathcal{R}|^2$ term of the hologram amplitude transmittance yields a bright spot on axis in the reconstructed image. The inverse Fourier transform of the $|\hat{O}|^2$ term yields an intensity that is the magnitude squared of the autocorrelation of the object amplitude transmittance, centered on the origin in the image plane, as predicted by the autocorrelation theorem of Fourier analysis. The inverse Fourier transform of the term $\mathcal{R}^* \hat{O}$ yields an image of the object, deflected upward away from the optical axis in the image plane. The term $\mathcal{R}\hat{O}^*$ yields a second image of the object in the image plane, this one deflected downwards away from the optical axis. If the angle of the reference wave is chosen large enough, the two images will not coincide with one another and will not be affected by the on-axis terms.

In principal, the Fourier transform frequency variables should be scaled by $1/\lambda f$, but we can neglect this factor provided that all focal lengths are identical.

Fourier Transform Hologram Simulation

We start by retrieving the same object that was used for the Gabor hologram simulations.

```
In[•]:= dataobject1 = dataobject1 = Import["Barchartdata.dat"];
        n = Length[dataobject1]

        1024
```

Because the use of the offset reference wave requires a much larger number of hologram and image pixels than the Gabor hologram, we downsample the object by a factor of 4 to make the image smaller. In addition, to make room for the two images that are deflected away from the optical axis, we need to enlarge the representation of the image plane, which we do by using the command `ArrayPad[array, n/2]`, which adds `n/2` pixels with value 0 to all sides of the reduced object.

```
In[•]:= dataobject2 = Downsample[dataobject1, 4];
        dataobject3 = ArrayPad[dataobject2, n / 2];
```

The width and height of this enlarged array are both as follows:

```
In[•]:= n2 = Length[dataobject3]

        1280
```

We can examine the reduced object with its zero padding.

```
In[•]:= Image[Reverse[dataobject3]]
```

Out[ ]=

Next we define the reference point source in the object plane. First create an **n2** × **n2** array of zeros, then set one of the elements of that array to a large number.

```
In[ ]:= ref = Parallelize[Table[0, {i, 1, n2}, {j, 1, n2}]];
ref[[250, 250]] = 100 000;
Dimensions[ref]
```

```
{1280, 1280}
```

We have chosen the reference point to be at coordinates (250, 250) and have set its amplitude to be 100,000. Now specify the intensity incident on the hologram plane and treat it as the complex amplitude transmitted by the hologram during reconstruction. The object and the reference point are assumed to be coplanar, and the hologram incident intensity is found by means of a Fourier transform.

```
In[ ]:= holointensity = Abs[dft[dataobject3 + ref]]^2;
```

The image produced by the hologram is found by treating the intensity recorded at the hologram plane as the amplitude distribution transmitted by the hologram during reconstruction and inverse Fourier transforming that amplitude distribution.

```
In[ ]:= imageintensity = Abs[dft[holointensity]]^2;
```

Finally, plot the intensity distribution in the image plane. The calculation required for this plot can take a number of minutes, depending on your particular computer.

```
In[ ]:= Image[
  Reverse[RotateRight[10^(-22) * imageintensity, {n2 / 2, n2 / 2}]]]
```

Out[ ]=

Occasionally, it is necessary to multiply the data array to be displayed by a constant (in this case 10^{-21}) to bring it into the proper range for `Image[]` to produce the best display.

From this image we see that the use of an angular-offset reference wave has separated the twin images from one another and from the terms near the optical axis. The twin images are seen to be mirror reflections of one another about the optical axis, and to lie in a common image plane, a characteristic of the Fourier hologram. The use of a very strong reference wave has resulted in twin images that are much brighter than the light near the optical axis.

9.3 Leith–Upatnieks Holography – The Fresnel Hologram

The Fresnel hologram does away with the Fourier transforming lenses, replacing them with free-space propagation. The recording and reconstruction geometries for this type of hologram are shown below.

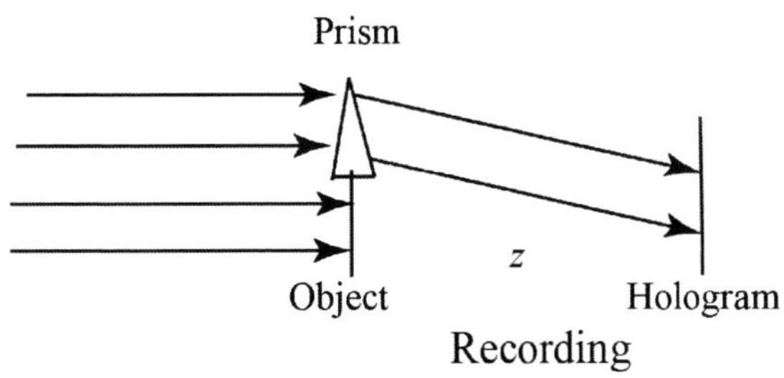

In[ ]:=

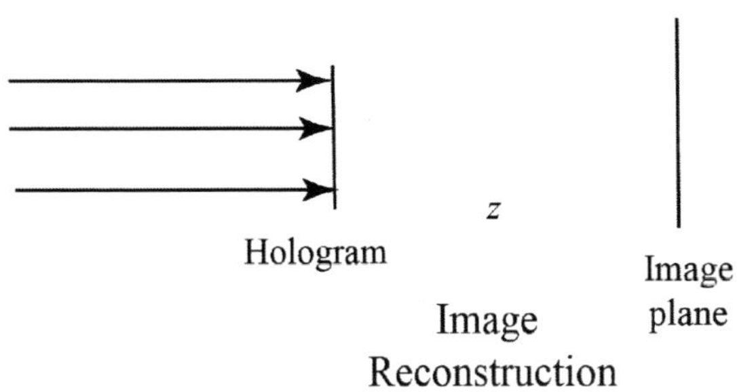

Thus, during recording of the hologram, we must propagate the light amplitude transmitted by the object over the distance z to the hologram recording plane, and there add the amplitude of the plane reference wave arriving at an angle. The hologram detects the intensity of the interference between these two waves. For reconstruction of the image, the intensity recorded must be converted into light amplitude transmitted by the hologram, following which propagation of the amplitude over distance z brings us to the plane where one of the twin images is in focus and the other is out of focus. Alternatively, we can back-propagate distance $-z$ to bring the other twin image into focus and the former image out of focus. The image intensity is detected in the image plane.

The Object and its Spectrum

For a starting image, we use the same image as was used in the previous sections and the same parameter **n**.

```
In[ ]:= dataobject1 = Import["Barchartdata.dat"];
        n = Length[dataobject1]

        1024
```

As in the previous section, we make room for the off-axis twin images by downsampling the object and adding zero-padding.

```
In[ ]:= dataobject2 = Downsample[dataobject1, 4];
        dataobject3 = ArrayPad[dataobject2, n / 2];
        Image[Reverse[dataobject3]]
Out[ ]=
```

Check the width and height of this expanded object.

```
In[ ]:= Dimensions[dataobject3]
        n2 = %[[1]];

        {1280, 1280}
```

Next we Fourier transform the expanded object, for we will need its transform when propagating light using the Fresnel transfer function method.

```
In[ ]:= spectrumampl = RotateRight[dft[dataobject3], {n2 / 2, n2 / 2}];
        Image[.003 * Reverse[Abs[spectrumampl]]]
```

Out[ ]=

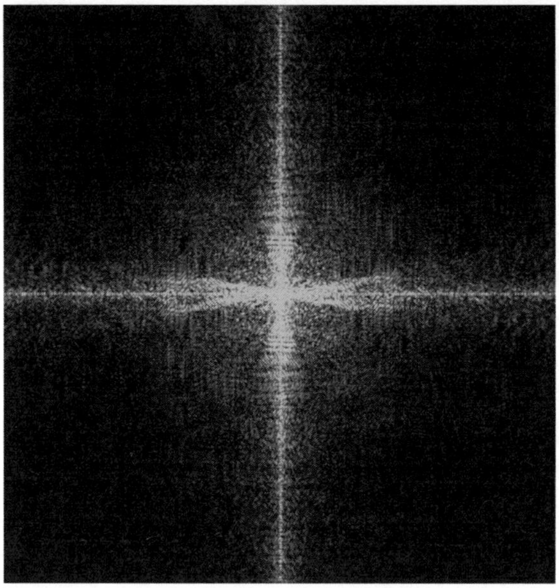

The Reference Wave

Now we construct the plane reference wave that is incident on the hologram, propagating at an angle to the optical axis. We construct this plane wave by first specifying its spectrum as a single point in a spectral array, and then Fourier transforming that array to produce the inclined plane wave. The angle of the arriving plane wave is determined by which spectral point we choose, and we can specify the strength of the plane wave by choosing the amplitude of that spectral point.

```
In[ ]:= preref = Parallelize[Table[0, {i, 1, n2}, {j, 1, n2}]];
        preref[[n2 / 4, n2 / 4]] = 1000;
        refampl = RotateRight[dft[preref], {n2 / 2, n2 / 2}];
```

Note that **preref** is an array of zeros, from which we choose one point to be equal to 1000. The reference wave at the hologram is represented by **refampl**.

The Transfer Function

Next we specify the transfer function that will be used in the calculation of the effects of propagation. We use the same transfer function as was used in the case of the Gabor hologram, but this time with parameters **n2** = 1280, Q = **q** = 20, and N_F = **nf** = 1. The real part of the complex-valued transfer function is displayed.

```
In[ ]:= q = 20; nf = 1;
        transferfunc = Parallelize[
           Table[Exp[-I * π / (4 * q^2 * nf) * ((k - n2 / 2) ^2 + (p - n2 / 2) ^2)],
            {k, 1, n2}, {p, 1, n2}]];
        ImageAdjust[Image[Re[transferfunc]]]
```

Out[ ]=

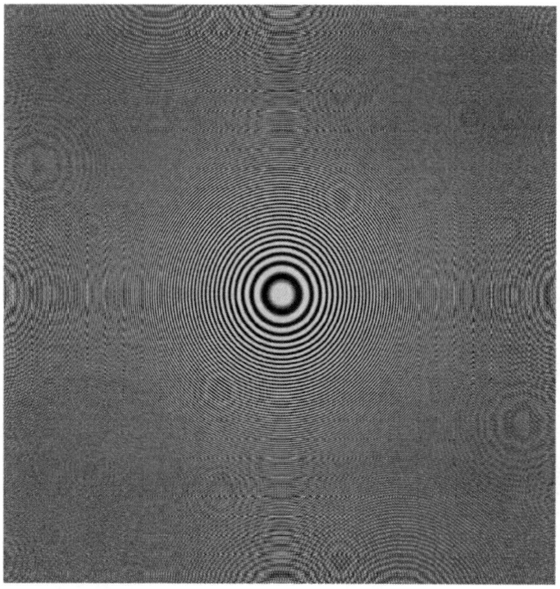

There are some weak signs of aliasing at the edges of the array; however, they lie in areas where there is very little spectral power of the object and therefore have only a minor effect on the image.

Detecting the Hologram

Next we calculate the intensity incident on the hologram recording plane, using the Fresnel transfer function method and including the plane reference wave.

```
In[ ]:= holoampl = idft[spectrumampl * transferfunc] + refampl;
holointens = Abs[holoampl]^2;
```

The amplitude transmittance of the hologram should be proportional to the intensity incident during detection of the hologram.

```
In[ ]:= holotransmittance = holointens - Mean[holointens];
```

We have subtracted the mean value of the hologram amplitude transmittance to reduce the light at the center of the image plane that the mean value would cause.

Calculating and Displaying the Real Image

The final steps are to propagate the fields away from the hologram plane using the same Fresnel number as was used for propagation from the object to the hologram, and to take the squared magnitude (intensity) of the field to observe the image. Note that the reconstruction beam illuminating the hologram is propagating normal to the hologram and from left to right. The code that follows calculates the in-focus real image.

```
In[ ]:= holoampl = idft[spectrumampl * transferfunc] + refampl;
holointens = Abs[holoampl]^2;
holotransmittance = holointens - Mean[holointens];
imageampl = dft[dft[holotransmittance] * Conjugate[transferfunc]];
imageintens = RotateRight[Abs[imageampl]^2, {n2 / 2, n2 / 2}];
Image[Reverse[10^(-18) * imageintens]]
```

Out[ ]=

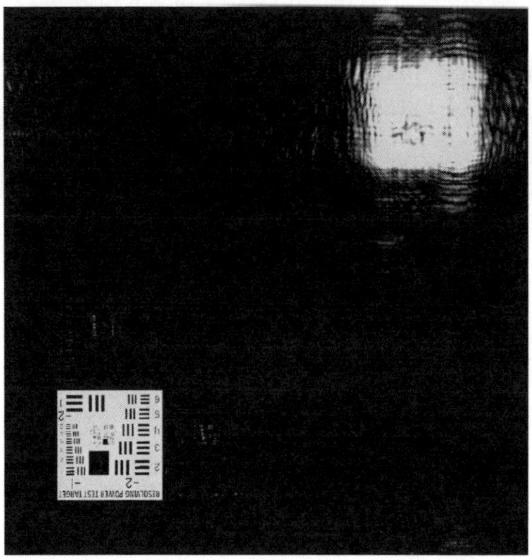

As you can see, for the Fresnel hologram, twin images are generated, but they lie in different planes; in this case, the real image appears inverted and in-focus and the virtual image out-of-focus. Because of the strong reference wave, the images are much brighter than the on-axis terms, which don't appear.

Calculating and Displaying the Virtual Image

It is possible to bring the virtual image into focus and the real image above out of focus. Normally, to view the virtual image the observer looks into the hologram from the left, and the eye forms an image from the diverging light from that image on the retina. An alternative approach, which we take here, is to illuminate the hologram with a normally incident plane wave traveling right to left, rather than left to right as in the case of the real image. The code below takes account of these conjugations and calculates the resulting image.

In[ ]:=
```
imageampl = idft[dft[holotransmittance] * transferfunc];
imageintens = Abs[imageampl] ^ 2;
ImageAdjust[
 Image[Reverse[RotateRight[10 ^ (-6) * imageintens, {n2 / 2, n2 / 2}]]],
 {0, 1}]
```

Out[ ]=

Simulating the Fresnel hologram is a more difficult task than simulating the Gabor hologram. First, one must pad the object with a sufficient number of zeros to allow room for the twin images, one in-focus, the other out-of-focus, and for the central components in the image plane. The separation of the two images is determined by both the spatial frequency of the reference term and the Fresnel number of the propagation transfer function. Thus, to achieve a satisfactory result, these three parameters, i.e., the padded field extent, the reference frequency, and the Fresnel number, must be carefully chosen. It is important to remember that the Fresnel number representing the propagation geometry does not correspond to the distance of propagation alone, but rather to the ratio of half the object boundary dimension squared to the propagation distance z (the wavelength λ is assumed to be fixed).

Why Not Use the Fresnel Transform Approach for Propagation?

We have used the Fresnel transfer function approach for simulating propagation in the examples in this chapter. When calculating simple diffraction patterns, this approach is generally preferred for large Fresnel numbers (> 1), while the Fresnel transform method is preferred for small Fresnel numbers (< 1). However, when calculating holograms that have free-space propagation, a problem is encountered with the Fresnel transform approach that is worth mentioning here.

The Fresnel transform approach calculates the field incident on the hologram plane $\mathbf{U}_H(p, q)$ according to the discrete equation

$$\mathbf{U}_H(p,q) = \exp\left[i\pi \frac{1}{4\,Q^2\,N_F}\left((p-N/2)^2+(q-N/2)^2\right)\right]\times$$

$$\sum_{n=0}^{N-1}\sum_{m=0}^{N-1}\mathbf{U}_o(n,m)\exp\left[i\pi\frac{4\,Q^2\,N_F}{N^2}\left((n-N/2)^2+(m-N/2)^2\right)\right]\times\exp\left[-i\frac{2\pi}{N}(p\,n+q\,m)\right].$$

Here $\mathbf{U}_H$ is the complex field incident on the hologram plane and $\mathbf{U}_o$ is the complex field transmitted by the object. Q is a factor that can be changed to avoid aliasing in the quadratic-phase exponential terms. N_F represents the Fresnel number appropriate for propagation from the object to the hologram plane. We have neglected a scaling factor and a complex constant that depends only on the distance z.

When calculating the intensity of a diffraction pattern of the object at some distance z, we can neglect the pre-multiplying exponential factor because it does not affect the intensity of the diffraction pattern. However, when simulating recording of a hologram, the pre-multiplying exponential factor must be retained because the entire complex field from the object interferes with the reference wave at the hologram plane. This factor and its conjugate in effect supply a lens-like focusing power that brings the twin images to focus in different planes. Thus, there are two quadratic-phase exponential factors to deal with, one the pre-multiplying factor and the other the first exponential factor that multiplies the object amplitude transmittance after the summation signs.

The difficulty with this approach becomes evident when we examine the magnitudes of the amplitude multipliers in each of the quadratic-phase exponentials. In the pre-multiplication exponential, the amplitude factor of concern is $\frac{1}{4\,Q^2\,N_F}$, while in the first exponential under the summation signs, the factor is $\frac{4\,N_F\,Q^2}{N^2}$.

Now consider how each of these factors might affect aliasing of each of the sampled quadratic-phase exponential terms. A large Fresnel number N_F decreases aliasing in the pre-multiplication factor but increases aliasing in the first exponential under the summation signs. A large Q decreases aliasing for the pre-multiplication factor but increases aliasing for the factor under the summation signs. Thus, there exist counter-requirements on N_F and Q, and achieving choices for these two factors that avoid aliasing in both exponential terms is challenging.

The approach needed to use the Fresnel transform approach would be the following. First choose $Q^2 N_F$ large enough to avoid aliasing in the pre-multiplication factor. Then choose N large enough to overcome the large $Q^2 N_F$ and thereby avoid aliasing in the quadratic-phase factor after the summation signs.

Note that when the Fresnel transfer function approach is used, there is only one quadratic phase exponential to deal with (i.e., the transfer function), so the problem of selecting parameters becomes somewhat simpler, although care must still be taken not to introduce aliasing that overlaps with the object spectrum.

9.4 Phase-Stepping Holography

Over the many years since its invention, holography has gone through many innovative changes and improvements. In this section, one of those innovations will be considered. In 1997, I. Yamaguchi and T. Zang demonstrated a technique that has come to be known as phase-stepping digital holography [13], based on earlier work on digital interferometry by several different researchers (see, for example, Ref. [14]). Also see Ref. [15] by J.C. Wyant for a more detailed discussion of this technique and its history. It is this technique that we consider here.

For this technique, our goal is to measure the amplitude and phase of a scalar object wave-field represented by $a(x, y)\, e^{i\,\theta(x,\,y)}$ impinging on a detector [note that $a(x, y)$ is a real and non-negative function]. To accomplish this goal, we interfere the object wave with an on-axis planar wavefront $A_0 e^{i\,\phi}$, where A_0 and ϕ are the constant amplitude and phase of a normally incident reference wave at the detector. The intensity measured by the detector when the reference-wave phase is ϕ is

$$I_\phi(x, y) = \left|A_0\, e^{i\,\phi} + a(x, y)\, e^{i\,\theta(x,y)}\right|^2 = A_0^{\,2} + a(x, y)^2 + 2\, A_0\, a(x, y) \cos[\theta(x, y) - \phi].$$

First we measure the amplitude of the object wave $a(x, y)$ by blocking the reference wave and allowing only the intensity of the object wave $a(x, y)^2$ to fall on the detector. A positive square root of this measured function then tells us the real and non-negative distribution $a(x, y)$.

To find the phase $\theta(x, y)$ of the object wave, we make four measurements, changing the phase ϕ of the reference wave by $\pi/2$ between measurements. The four detected intensities are

$$\begin{aligned}
I_0(x, y) &= A_0^{\,2} + a(x, y)^2 + 2\, A_0\, a(x, y) \cos[\theta(x, y)], \\
I_{\pi/2}(x, y) &= A_0^{\,2} + a(x, y)^2 + 2\, A_0\, a(x, y) \cos[\theta(x, y) - \pi/2], \\
I_\pi(x, y) &= A_0^{\,2} + a(x, y)^2 + 2\, A_0\, a(x, y) \cos[\theta(x, y) - \pi], \\
I_{3\,\pi/2}(x, y) &= A_0^{\,2} + a(x, y)^2 + 2\, A_0\, a(x, y) \cos[\theta(x, y) - 3\,\pi/2].
\end{aligned}$$

It is now possible to find the phase distribution $\theta(x, y)$ from the following formula:

$$\theta(x, y) = \arctan\left(\frac{I_{\pi/2} - I_{3\,\pi/2}}{I_0 - I_\pi}\right).$$

To illustrate this technique using *Mathematica*®, we choose to apply this method to finding the distribution of amplitude and phase in a Fourier transform of a simple function, a product of two rectangle functions rect(x) rect(y). Of course, we already know that the transform is sinc(ν_x) sinc(ν_y), but we will use this knowledge only as a check on our calculated results. Since the applications of this technique involve sampled data from detectors, we will use sampled data in our illustration.

First begin by clearing the variables and list names we will use:

```
In[ ]:= Clear[g, g1, G, a, i, n, ϕ, θ, a0]
```

We set the parameter **n** to 64, but we create an array size that is 257 × 257 pixels by using a step increment of 1/4 in the table:

```
In[*]:= n = 64;
        g = Parallelize[Table[rect[x / 2] * rect[y / 2],
            {x, -n / 2, n / 2, 1 / 4}, {y, -n / 2, n / 2, 1 / 4}]];
```

Now a subtle point: The created list **g** has the rectangle functions centered in the array. To avoid linear phase shifts implied by the shift theorem, we must move the rectangles so that their centers lie at indices (1, 1), with periodic wrap-around. Otherwise, the linear phase shift will obscure the phases we wish to find.

```
In[*]:= g1 = RotateLeft[g, {2 * n, 2 * n}];
```

Now we apply a Fourier transform and, purely for convenience, center the transform at indices (**2n**, **2n**), naming the complex-valued transform list $\mathcal{G}$.

```
In[*]:= 𝒢 = RotateRigh [dft[g1], { 2*n,2 *n}];
```

The modulus of the Fourier transform is the list **a**:

```
In[*]:= a = Abs[𝒢];
```

We next plot the modulus of $\mathcal{G}$:

```
In[*]:= ListPlot3D[a, PlotRange → All]
```

Out[*]=

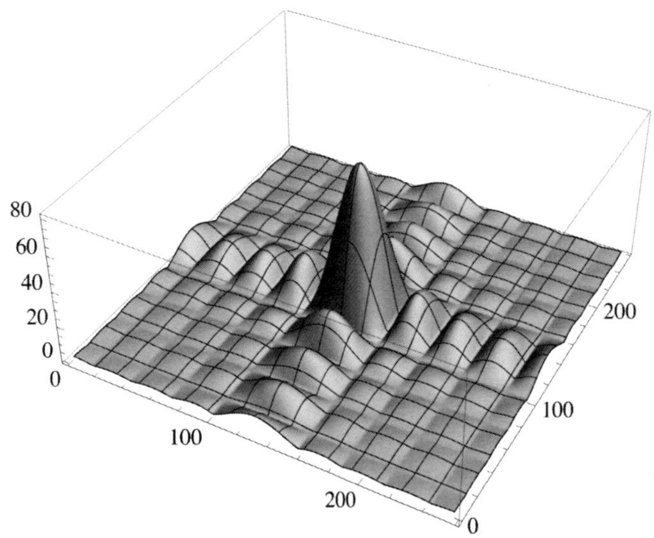

We now know the modulus of the field, and must turn to finding the phase distribution. Define a general intensity function **i[a0,ϕ]** representing the interference between a reference wave of uniform modulus **a0** and phase **ϕ** and an object wave represented by the Fourier transform $\mathcal{G}$.

```
In[*]:= i[a0_, ϕ_] := Abs[a0 * Exp[I * ϕ] + 𝒢] ^2
```

Now introduce the equation for finding the object phase θ. We set the reference wave modulus **a0** to twice the maximum value of the array **a**:

```
In[*]:= a0 = 2 * Max[Abs[𝒢]];
```

```
In[ ]:= Quiet[
        θ=ArcTan[(i[a0,0]-i[a0,Pi]),( i[a0,Pi/2]-i[a0,3*Pi/2])-10^(-6)]];
```

Note that we have subtracted a factor of 10^{-6} from the ratio of intensity differences. This has been done to avoid the effects of numerical imprecision, which at a few points drives the value of the calculated phase above 0, resulting in the wrap-around of the phase to value -2π when it should be 0. Finally, we plot the phase function we have just found:

```
In[ ]:= ListPlot3D[θ, PlotRange → All]
```

In[ ]:=

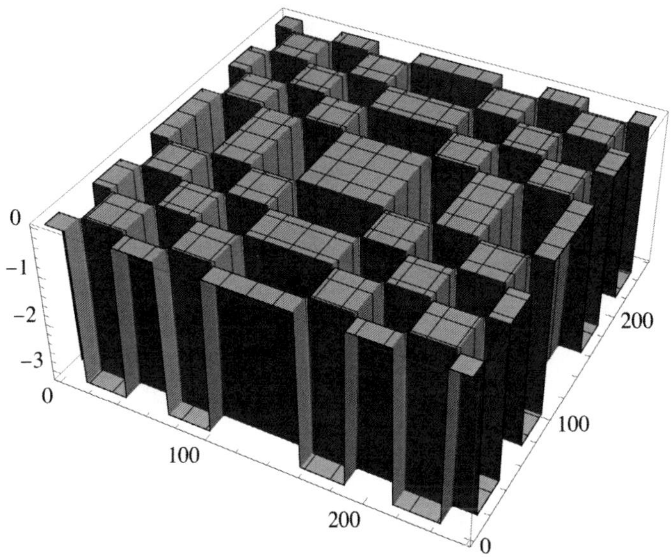

We see the phase alternating between 0 and $-\pi$, i.e., the phase of the sidelobes alternating between positive and negative values in adjacent lobes of the sinc function, as should be expected from our knowledge of the known result. We have found the entire amplitude and phase functions of the Fourier transform using four interferometric measurements, which is the goal of holography. It is also possible to use three measurements at (0, $2\pi/3$, $4\pi/3$) reference phase shifts, with a different arctan equation for θ. Four or more phase measurements are usually used to improve the signal-to-noise ratio. From the amplitude and phase functions in the diffraction pattern that we have found, we can reconstruct an image of the original function that gave rise to these results.

10. Spatial Filtering With Coherent Optics

We have seen in Chapter 7 that the transmittance through the central focal plane of a $4f$ system affects the appearance of the coherent image produced by a given object. This property allows us to analyze coherent optical imaging systems and calculate the images they produce. A closely related concept is that of spatial filtering by means of a coherent optical system, for which the goal is not to predict the image structure, but rather to manipulate the image structure in some desirable way. In this final chapter, we illustrate coherent spatial filtering with simulations using *Mathematica*®. For a much more detailed discussion of coherent optical filtering, see Ref. [4], Chapter 10.

10.1 An Object to be Filtered

For the object, we continue to use the resolution chart used in several previous chapters. We first import that image.

```
In[ ]:= dataobject1 = Import["Barchartdata.dat"];
        Image[Reverse[dataobject1]]
```

Out[ ]=

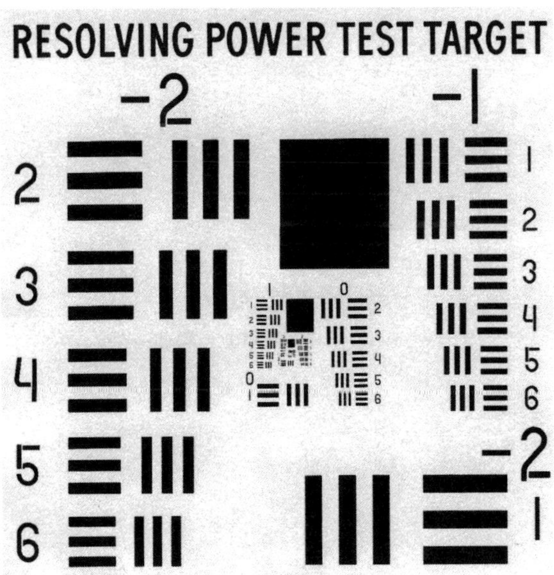

The image is a 1024×1024 data array.

```
In[ ]:= n = Length[dataobject1]
```

```
1024
```

The spatial filtering system is again a $4f$ system, and we imagine that the input to the system is a transparency with an amplitude transmittance proportional to the above image.

Because we will be manipulating the spectrum of this object in the focal plane, we display the intensity of that spectrum again.

```
In[ ]:= spectrum1 = RotateRight[Fourier[dataobject1], {n / 2, n / 2}];
        Image[Reverse[Abs[spectrum1] ^ 2]]
```

Out[ ]=

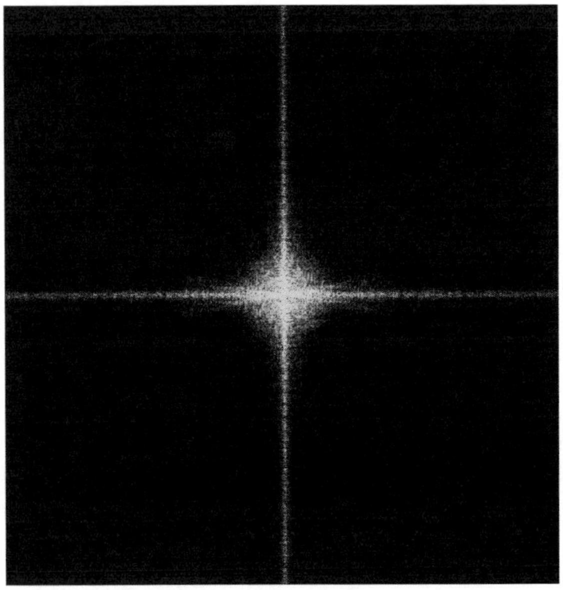

The strong and wide horizontal component of the spectrum arises from the many vertically oriented bars in the object, while the strong and tall vertically oriented component of the spectrum arises from the many horizontally oriented bars in the object. The small cloud about the origin of the spectrum arises in part from some of the characters in the title at the top of the object.

10.2 Filtering to Emphasize Horizontal or Vertical Bars

Our goal in this section is to demonstrate selective emphasis of horizontal bars or vertical bars in the image by spatial filtering. To emphasize horizontal bars and block vertical bars, we need a filter in the central focal plane of the $4f$ system that blocks only the horizontal stripe in the spectrum of the object, as the horizontal stripe arises from the vertical bars. Such a filter can be made by a thin opaque mask in the focal plane that blocks light in the horizontal stripe in the spectrum. The amplitude transmittance of that Fourier-plane filter can be defined by the following code:

```
In[ ]:= blockhor :=
          Parallelize[Table[(1 - rect[(j - n / 2) / 30]), {i, 1, n}, {j, 1, n}]];
```

Analogously, a filter to emphasize the vertical bars and block horizontal bars of the object requires a filter that blocks only the vertical stripe in the spectrum, as it arises from the horizontal bars in the object.

```
In[ ]:= blockver := Parallelize[Table[(1 - rect[(i - n / 2) / 30
              ]), {i, 1, n}, {j, 1, n}]];
```

Images showing the light intensity transmitted through the focal plane are shown below.

```
In[ ]:= p1 = Image[Reverse[Abs[blockhor * spectrum1] ^2]];
        p2 = Image[Reverse[Abs[blockver * spectrum1] ^2]];
        GraphicsRow[{p2, p1}, ImageSize → 400]
```

Out[ ]=

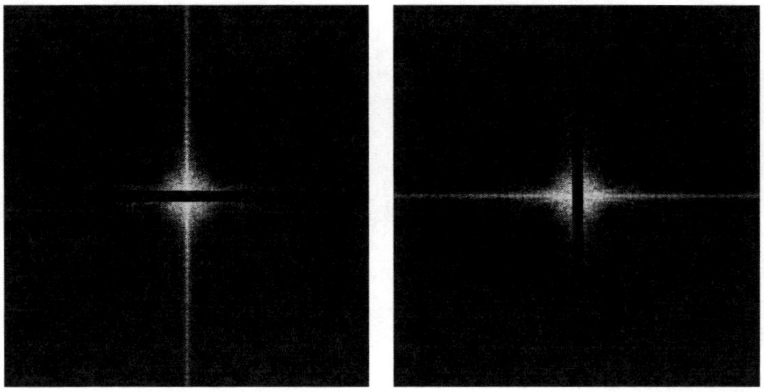

The horizontal-blocking stripe is on the left, and the vertical-blocking stripe is on the right. Now define the light amplitude distributions in the image plane of the $4f$ system in the two cases.

```
In[ ]:= (* Vertical image amplitude and an image *)
vertical = dft[spectrum1 * blockhor];
q1 = Image[5 * Reverse[Abs[vertical]^2] / Max[Abs[vertical]^2]];
```

```
In[ ]:= (* Horizontal image amplitude and an image *)
horizontal = dft[spectrum1 * blockver];
q2 =
  Image[5 * Reverse[Abs[horizontal]^2] / Max[Abs[horizontal]^2]];
```

We have multiplied the images by a factor of 5 to increase their brightness. The images are now shown side-by-side.

```
In[ ]:= GraphicsRow[{q1, q2}, ImageSize → 400]
```

Out[ ]=

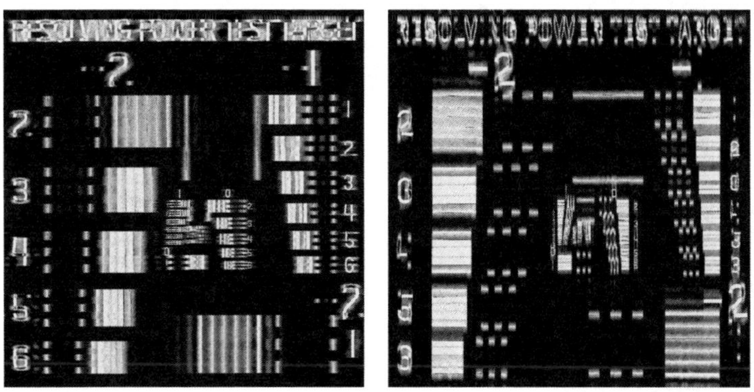

The result of blocking the spectrum with a vertical opaque stripe is shown on the left, and blocking with a horizontal opaque stripe is on the right. The blocking of horizontal structures is clear on the left, and blocking of vertical structures is clear on the right. On the left, notice that horizontal bars have ends with small vertical segments, and those segments show up clearly. The situation is similar for the small horizontal segments at the ends of vertical stripes in the right-hand image. Notice also that the title at the top of each image has more vertical lines than horizontal lines.

10.3 Edge Enhancement

A second simulation illustrates the effect of blocking the central region of the spectrum. Let the Fourier plane blocking filter be defined by

```
In[•]:= central := Parallelize [Table[
         (1-circ[Sqrt[(i-n/2)^2 + (j-n/2)^2]/64]),{ i,1,n}, {j,1,n}]];
```

A plot of the intensity transmitted through the focal plane follows:

```
In[•]:= Image[Reverse[Abs[central * spectrum1] ^2], ImageSize → 300]
```

Out[•]=

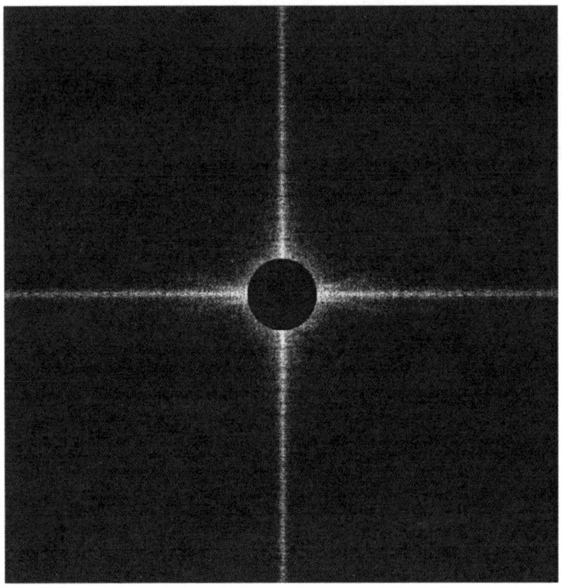

Now calculate and show the image intensity that results from blocking this central region of the spectrum.

```
In[•]:= edges = dft[spectrum1 * central];
        Image[9 * Reverse[Abs[edges] ^2] / Max[Abs[edges] ^2]]
```

Out[•]=

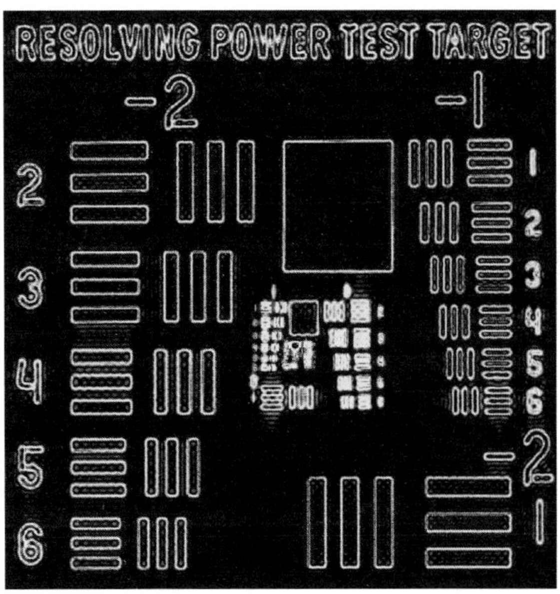

The enhancement of edges is clear. Also note that the smallest bar triplets have their edges enhanced the most, due to the fact that most of their spectral energy lies well outside the central blocking filter.

10.4 Restoring Resolution in an Incoherent Blurred Image

In this section, we imagine that an incoherent imaging system forms an intensity image that has been blurred by a Gaussian point-spread function (PSF). The Gaussian blur may be an approximation to any of a number of aberrations, including the blur introduced by atmospheric refractive-index irregularities.

Creating the Blurred Image

For our simulations, we use the fact that the Fourier transform of a Gaussian function is itself a Gaussian function, and that the autocorrelation function (necessary to find the OTF) of a Gaussian function is also a Gaussian function. Thus, the OTF of the imaging system has a Gaussian shape, and we represent the OTF with the function gaus(r) = $e^{-\pi r^2}$, where r stands for radius. We choose the $1/e$ radius of the Gaussian OTF to be $8/\sqrt{\pi}$. *The Mathematica* representation of the OTF is

```
In[ ]:= OTF = Parallelize[Table[gaus[Sqrt[((i - n / 2) ^2 + (j - n / 2) ^2)] / 64],
          {i, 1, n}, {j, 1, n}]];
```

A plot of this function follows.

```
In[ ]:= ListPlot3D[OTF]
Out[ ]=
```

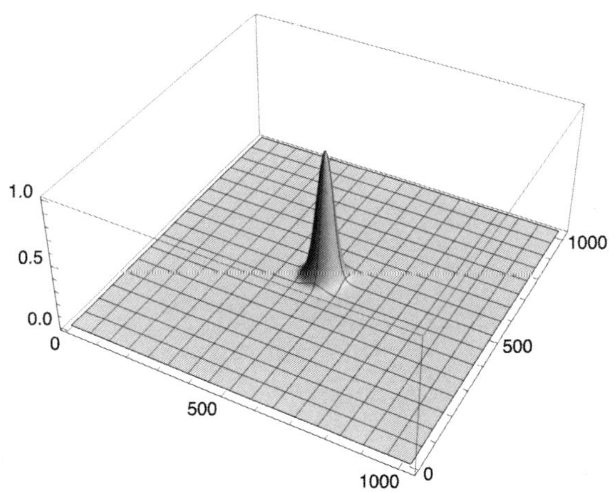

To blur the image, we multiply the Fourier spectrum of the object by the OTF and inverse Fourier transform the result. The blurred image intensity is shown below.

```
In[ ]:= blurred = Abs[idft[spectrum1 * OTF]];
        ImageAdjust[Image[Reverse[Abs[blurred], 2]], {0, 1000}]
```

Out[•]=

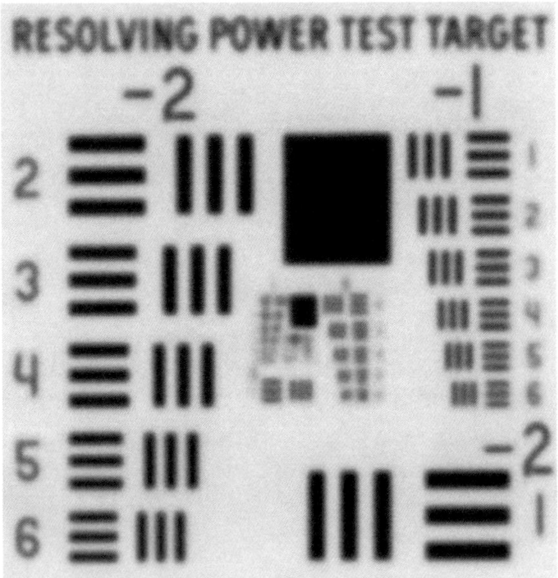

The blur has destroyed our ability to resolve the fine structure below the large black square. This image is the image we wish to improve by filtering in a coherent 4*f* system.

10.5 An Inverse Filter

We assume that the input to the coherent optical system is a transparency with amplitude transmittance proportional to the blurred image intensity above. We can manipulate this image by means of a filter in the focal plane of the 4*f* coherent optical system.

To illustrate a point, we choose a filter that is the inverse of the blur OTF.

```
In[•]:= invfilter := Table[
          1 / gaus[Sqrt[(i - n / 2) ^2 + (j - n / 2) ^2] / 64], {i, 1, n}, {j, 1, n}];
```

The filtered image intensity can now be calculated and displayed. Note that the command **Reverse[deblurred, 2]** reverses the order of the columns of the image, a necessary step to properly orient the image. The spectrum of the unblurred image is multiplied by the OTF, and that product is then multiplied by the inverse filter. This new spectrum is then subjected to an inverse DFT, and the image intensity is calculated.

```
In[•]:= deblurred = Abs[idft[(spectrum1 * OTF) * invfilter]] ^2;
```

```
In[•]:= ImageAdjust[Image[Reverse[deblurred, 2] / Max[deblurred]]]
```

Out[ ]=

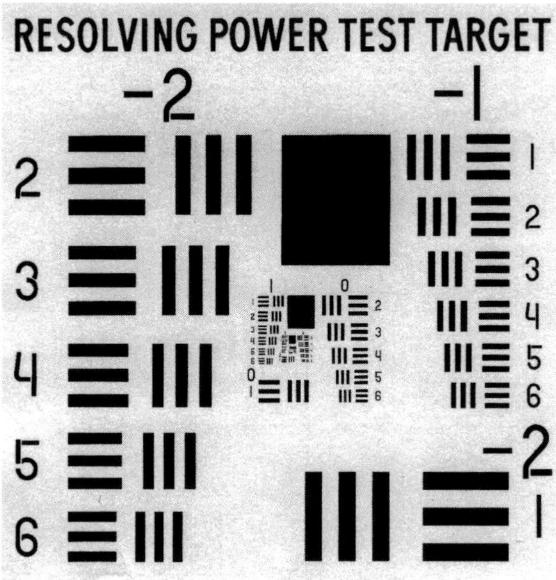

The inverse filter has totally restored the image to essentially its unblurred state. However, we have forgotten something extremely important in our model. When the original blurred image was detected in the incoherent system, the detected data was accompanied by additive detector noise. We now include that noise, modeling it as having zero mean and Gaussian or normal statistics, independent at all pixels. We choose an exceedingly small standard deviation of that noise, 10^{-100}, to illustrate our point.

In[ ]:=
```
noise = Parallelize[Table[RandomVariate[
     NormalDistribution[0, 10^(-100)]], {i, 1, n}, {j, 1, n}]];
```

We form an image of the noisy blurred image. The noise is too weak to be seen.

In[ ]:=
```
blurred1 = idft[(spectrum1 * OTF) + dft[noise]];
ImageAdjust[Image[Reverse[Abs[blurred1], 2]], {0, 1200}]
```

Out[ ]=

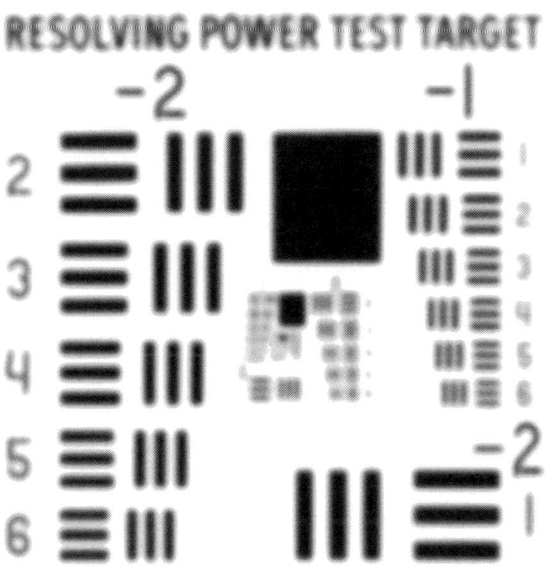

Now we use the same inverse filter used previously, this time for the blurred image with noise, with disastrous results. Because of the extremely large value of the inverse filter at high frequencies, we have had to multiply the image by a very small number to display it.

```
In[ ]:= Image[Reverse[10^(-72)*Abs[idft[((spectrum1*OTF)+dft[noise])
        *invfilter]],2]]
```

Out[ ]=

The resulting image does not have any features of the original object; noise has obliterated the image of the object. This has happened because the inverse filter has boosted the noise spectrum in the outer parts of the spectral array by a very large amount, while not boosting the signal spectrum by a comparable amount since it is mostly confined to a small region about the origin. Note that the value of the inverse filter transfer function at indices (**n**, **n**) is approximately 10^{174}, and it is even larger in the diagonal corners of the array.

We conclude that the inverse filter does not work in practical situations, so a modification of the restoration filter is needed.

10.6 The Wiener Filter

A restoration filter designed to be less sensitive to noise than the inverse filter was published by Norbert Wiener in 1949 (Ref. [16]). The form of the transfer function $\mathcal{W}$ of this filter is

$$\mathcal{W}(\nu_x, \nu_y) = \frac{\mathbf{B}^*(\nu_x, \nu_y)}{|\mathbf{B}(\nu_x, \nu_y)|^2 + \frac{\Phi_n(\nu_x, \nu_y)}{\Phi_s(\nu_x, \nu_y)}},$$

where $\mathbf{B}(\nu_x, \nu_y)$ represents the transfer function of the blur, and $\Phi_n(\nu_x, \nu_y)$ is the power spectral density of the noise. In this approach, the unblurred image is treated as a random process with power spectral density given by $\Phi_s(\nu_x, \nu_y)$. Note that at all frequencies where $\Phi_s(\nu_x, \nu_y) \gg \Phi_n(\nu_x, \nu_y)$, the restoration filter becomes close to an inverse filter, i.e., $\mathcal{W}(\nu_x, \nu_y) \to 1/\mathbf{B}(\nu_x, \nu_y)$. However, at frequencies where the noise power spectral density is comparable with or greater than the the signal power spectral density, the restoration filter introduces attenuation and does not amplify the noise.

The appropriateness of the use of this filter can be questioned because the original object is not a random process with a known power spectral density. Rather, it is an unknown single object. We cannot assume that we know the Fourier spectrum of the object, for if we did know it, we could calculate the unblurred image directly from that spectrum. What we will assume here as apriori knowledge is that the object is drawn from an ensemble of random objects consisting of both horizontal and vertical bars, with unknown arrangements and locations in the object field. As a reasonable assumption about the power spectral density of the collection of such objects, we assume that it consists of a uniform horizontal stripe along the ν_x axis and a uniform vertical stripe along the ν_y axis, both 30 pixels wide.

```
In[•]:= horiz = Parallelize[Table[rect[(i - n / 2) / 30], {i, 1, n}, {j, 1, n}]];
        vert = Parallelize[Table[rect[(j - n / 2) / 30], {i, 1, n}, {j, 1, n}]];
        psdSignal = horiz + vert + 10^(-5);
```

The small constant 10^{-5} is added to avoid later division by zero. The assumed power spectral density of the ensemble of possible objects is now shown.

```
In[•]:= Image[psdSignal, ImageSize → 200]
Out[•]=
```

The noise sequence is regenerated now with the same standard deviation that was previously used for the inverse filter.

```
In[•]:= noise1 = Parallelize[Table[RandomVariate[
              NormalDistribution[0, 10^(-100)]], {i, 1, n}, {j, 1, n}]];
```

The noise power spectral density can be found by first finding the autocorrelation function of the noise and then Fourier transforming that autocorrelation function. The last 1 in `ListCorrelate[]` below is a parameter that in effect limits the dimensions of the correlation function to the same as those of `noise1`. We calculate the power spectral density of the noise.

```
In[•]:= autocorrNoise = ListCorrelate[noise1, noise1, 1];
        psdNoise = Abs[Fourier[autocorrNoise]];
```

We now calculate and display the normalized Wiener filter transfer function amplitude transmittance.

```
In[•]:= Quiet[𝒲 = OTF / (Abs[OTF]^2 + (psdNoise / psdSignal))];
        ImageAdjust[Image[𝒲 / Max[𝒲], ImageSize → 300], {0, 1000}]
```

Out[ ]=

The bright circle near the edges of the display comes from the weak background we added to the signal power spectral density at all pixels in order to avoid division by zero. With the choice of noise standard deviation we have made, this filter is close to an inverse filter, which has its maxima at very high spatial frequencies (the maximum value of $\mathcal{W}$ is approximately 10^{95}). However, the transmitted spectrum remains significant within the circle along the ν_x and ν_y axes, and most of those frequencies are properly restored. Next we look at the image amplitude restored by the Wiener filter.

In[ ]:=
```
image = Abs[idft[(spectrum1 * OTF + dft[noise1]) * 𝒲]];
Image[Reverse[10^3 * image, 2]]
```

Out[ ]=

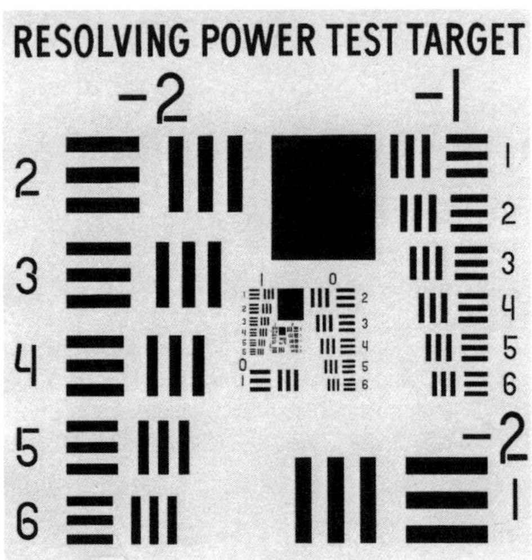

The image is nearly perfect, showing that the Wiener filter can properly restore the image when the noise standard deviation is as small as assumed above and the apriori information we assumed is built into the filter model. Now we put the various command lines together using a much stronger noise standard deviation of 10^{-3} instead of 10^{-100}. How will this affect the filter and the restored image? The image below shows the normalized transmittance of the Wiener filter.

In[ ]:=
```
Clear[autocorrNoise, psdNoise, image]
noise2 =
  Parallelize[Table[RandomVariate[NormalDistribution[0, 10^(-3)]],
    {i, 1, n}, {j, 1, n}]];
```

```
autocorrNoise = ListCorrelate[noise2, noise2, 1];
psdNoise = Abs[Fourier[autocorrNoise]];
Quiet[𝒲 = OTF / (Abs[OTF] ^2 + psdNoise / psdSignal)];
ImageAdjust[Image[𝒲 / Max[𝒲], ImageSize → 300], {0, 1000}]
```

Out[ ]=

Because of the stronger level of noise, significant amplitude transmittance extends only around the central part of the transfer function along the ν_x and ν_y axes. Finally, we show the image amplitude that results in this case.

```
In[ ]:= image = Abs[idft[(spectrum1 * OTF + dft[noise1]) * 𝒲]];
ImageAdjust[
 Image[Reverse[image / Max[image], 2], ImageSize → 300], {0, 0.5}]
```

Out[ ]=

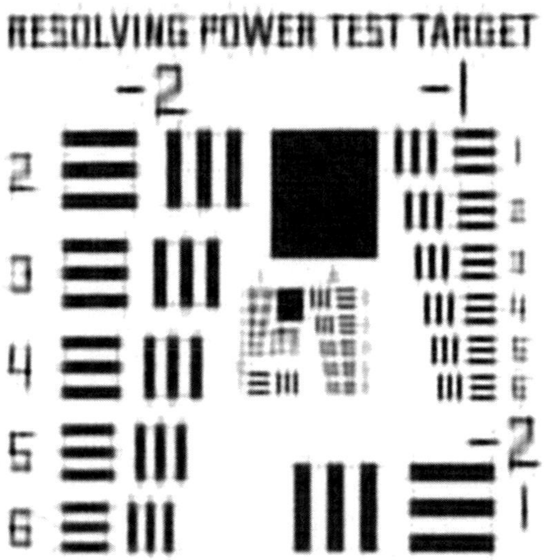

As can be seen, the higher level of noise has destroyed resolution for most of the smaller bar triplets, but the Wiener filter has prevented noise from dominating the result, as was the case for the inverse filter. Thus, the beauty of the Wiener filter is that it trades off resolution for freedom from disastrous noise.

Acknowledgment

I owe many thanks to Ms. Dara Burrows of SPIE Press for the consistency improvements she suggested. I have incorporated all of them, and the book has been significantly improved due to her efforts.

References

[1] J.W. Goodman, *Fourier Transforms Using* Mathematica®, SPIE Press, Bellingham, Washington (2020) [doi: 10.1117/3.2574956].

[2] R.N. Bracewell, *The Fourier Transform and Its Applications*, 2nd Edition, McGraw Hill, New York (1986).

[3] A. Papoulis, *The Fourier Integral and Its Applications*, McGraw Hill, New York (1962).

[4] J.W. Goodman, *Introduction to Fourier Optics*, 4th Edition, MacMillan Learning, New York (2017).

[5] G.C. Sherman, "Application of the convolution theorem to Rayleigh's integral formulas," *J. Opt. Soc. Am.* **57**(4), 546–547 (1967).

[6] J.R Fienup, "Fourier-Optics Imaging Analysis with ABCD Matrices: Tutorial," *J. Opt. Soc. Am. A* [in press].

[7] J.W. Goodman, *Statistical Optics*, 2nd Edition, J. Wiley & Sons, New York (2015).

[8] E. Wolf, *Introduction to the Theory of Coherence and Polarized Light*, Cambridge University Press, Cambridge, UK (2007).

[9] J.W. Goodman, *Simulating Speckle With* Mathematica®, SPIE Press, Bellingham, Washington (2022) [doi: 10.1117/3.2633988].

[10] J.W. Goodman, *Simulating Speckle With Python*, SPIE Press, Bellingham, Washington (2024) [doi: 10.1117/3.100010].

[11] D. Gabor, "A new microscope principle," *Nature* **161**(4098), 777–778 (1948).

[12] E.N. Leith and J. Upatnieks, "Reconstructed wavefronts and communication theory," *J. Opt. Soc. Am.* **52**(10), 1123–1139 (1962).

[13] I. Yamaguchi and T. Zhang, "Phase-shifting digital holography," *Opt. Lett.* **22**(16), 1268–1270 (1997).

[14] K. Creath, "Phase-shifting speckle interferometry," *Appl. Opt.* **24**(18), 3053–3058 (1985).

[15] J.C. Wyant, https://wp.optics.arizona.edu/jcwyant/wp-content/uploads/sites/13/2016/08/Phase-Shifting-Interferometry.nb_.pdf

[16] N. Wiener, *Extrapolation, Interpolation, and Smoothing of Stationary Time Series: With Engineering Applications*, MIT Press, Cambridge, Massachusetts (1949).